Matters
of
Spirit

F. Scott Scribner

Matters of Spirit

J. G. Fichte and the Technological Imagination

The Pennsylvania State University Press
University Park, Pennsylvania

Library of Congress Cataloging-in-Publication Data

Scribner, F. Scott.
Matters of spirit : J. G. Fichte and the technological
imagination / F. Scott Scribner.
    p.   cm.—(American and European philosophy)
"Earlier versions of this work were also tested in the form of
conference presentations at the North American Fichte
Society"—Acknowledgments.
Includes bibliographical references and index.
Summary: "An interpretation of the philosophy of J. G.
Fichte. Examines the impact of nineteenth-century
psychological techniques and technologies on the formation
of his theory of the imagination"—Provided by publisher.
ISBN 978-0-271-03621-2 (cloth : alk. paper)
ISBN 978-0-271-03475-1 (pbk : alk. paper)
 1. Fichte, Johann Gottlieb, 1762–1814.
 2. Imagination.
 3. Technology.
 I. Title.

B2849.I4S37   2010
193—dc22
2009036138

The Pennsylvania State University Press is a member of the
Association of American University Presses.
It is the policy of The Pennsylvania State University Press to
use acid-free paper. Publications on uncoated stock satisfy the
minimum requirements of American National Standard for
Information Sciences—Permanence of Paper for Printed
Library Material, ANSI z39.48-1992.

Contents

Acknowledgments

In the course of writing a book I incurred innumerable debts. As gifts they can never be repaid, but only acknowledged. I would first like to thank Max Pensky, Frederick Neuhouser, and Stephen David Ross, all of whom read an earlier version of this project. I would especially like to thank Allen Wood for his extensive critical commentary on an earlier draft. Although the project was always a bit too Continental in flavor for him, he helped me anyway. Thanks Allen! Earlier versions of this work were also tested in the form of conference presentations at the North American Fichte Society. I would especially like to thank the following members of the Society for their continued support and helpful commentary: Dan Breazeale, Tom Rockmore, Wayne Martin, Günter Zöller, Steven Holztel, Claude Piché, Michael Vater, and Robert Williams. To those who helped foster the best models of critical reasoning in the Continental style, I thank you: William Wurzer, Christopher Fynsk and William Haver. I would like to thank my colleagues at the University of Hartford for their support—especially Marcia Moen, David Goldenberg, Bill Stoll, Mari Firkatian, Anthony Rauche, and the entire Humanities Department. Since this is my first book, I also feel compelled to acknowledge my debt to some of my earliest teachers: Hugh Silverman, who was always generous with his time, and Richard Capobianco, who first helped shape my rebellion into wonder. And despite this great intellectual community, any mistakes within this book remain mine own alone.

Earlier versions of this work have appeared in edited collections. I would like to thank Rodopi Publishers for allowing me to use portions of the following two articles, which appear in heavily revised form as Chapters 4 and 7: "Falsification: On the Role of the Empirical in J. G. Fichte's Transcendental Method," in *Fichte, German Idealism and Early Romanticism (Fichte-Studien Supplementa)*, ed. Daniel Breazeale and Tom Rockmore (Amsterdam:

Rodopi, 2009), and *"Die 'Physicirung des Idealismus' im Tagebuch über den animalischen Magnetismus: Die letzte Wissenschaftslehre oder Das Ende des Idealismus?"* in *Fichte-Studien (Bd. 17–18): Die Spätphilosophie J. G. Fichtes,* ed. Wolfgang H. Schrader (Amsterdam: Rodopi, 2001). I would like to thank Northwestern University Press for allowing me to publish a revised version of the following article as Chapter 5: "The 'Subtle Matter' of Intersubjectivity in the *Grundlage des Naturrechts,*" in *New Essays on Fichte's Later Jena Wissenschaftslehre,* ed. Daniel Breazeale and Tom Rockmore (Evanston: Northwestern University Press, 2002). Finally, I would like to thank Ashgate Publishers for allowing me to publish as Chapter 6, a revised version of "The Aesthetics of Influence: Fichte's *Grundlage des Naturrechts* in View of Kant's Third Critique," in *Rights, Bodies, and Recognition,* ed. Daniel Breazeale and Tom Rockmore (Aldershot, U.K.: Ashgate, 2005).

As any self-respecting phenomenology teaches, intellectual debt is only half the picture. I would like to thank my parents for providing material support at key moments of this project and to Sharon (Mazzola) Snow, who provided the emotional warmth and lightness that made this project possible at all. Like oxygen on the mountain top, she's more Nietzschean than she'll ever know. I would like to thank my broader family for their support, especially Evelyn Eisenhardt and Barbara Hindley, who cared enough to ask me to explain my project in accessible terms. Finally I would like to dedicate this book to two people: to my dear friend Anthony Hakian, who knew the pain of solitude and answered the ultimate of philosophical questions in his own way (1968–1996), and to my grandfather, Charles E. Eisenhardt, who, by example, taught me how to love a book (1907–2006).

Abbreviations

All translations of Fichte's writings are mine except where otherwise noted. I generally follow Daniel Breazeale's translations where I used them, but substantially corrected other English translations.

Works by Fichte

EPW   *Fichte: Early Philosophical Writings*. Translated and edited by Daniel Breazeale. Ithaca: Cornell University Press, 1988.

FW   *Fichtes Werke*. 11 vols. Edited by I. H. Fichte. 1834–35. Reprint. Berlin: Walter de Gruyter, 1971.

GA   *J. G. Fichte Gesamtausgabe der Bayerishen Akademie der Wissenschaften*. Edited by Reinhardt Lauth, Hans Jacob, and Hans Gliwitsky. Stuttgart and Bad Cannstatt: Friedrich Frommann, 1964– . Citations by volume, part, and page.

GNR   *Grundlage des Naturrechts*. In *FW* 3:1–385.

IWL   *J. G. Fichte: Introductions to the Wissenschaftslehre and Other Writings*. Translated and edited by Daniel Breazeale. Indianapolis: Hackett, 1994.

SK   *The Science of Knowledge*. Translated and edited by Peter Heath and John Lachs. Cambridge: Cambridge University Press, 1982.

SLP   "On the Spirit and the Letter of Philosophy, in a Series of Letters." Translated by Elizabeth Rubenstein. In *German Aesthetic and Literary Criticism: Kant, Fichte, Schelling, Schopenhauer, Hegel*, ed. David Simpson, 74–93. Cambridge: Cambridge University Press, 1984.

TaM   "Tagebüch über den animalischen Magnetismus" appears in *FW*

11. This citation, however, refers to the draft of the version edited
by Ives Radrizzani for *J. G. Fichte: Gesamtausgabe der Bayerischen
Akademie der Wissenschaften* (forthcoming).

WLnm   *Fichte: Foundations of Transcendental Philosophy (Wissenschaftslehre)
Nova Methodo (1796/99).* Translated and edited by Daniel
Breazeale. Ithaca: Cornell University Press, 1992.

Other Works

AS   Schelling, F. W. J. *Ausgewählte Schriften.* Edited by Manfred Frank.
6 vols. Frankfurt: Suhrkamp Verlag, 1985.

CJ   ———. *Critique of Judgment.* Translated by Werner Pluhar. Indianapolis: Hackett, 1987.

CPR   ———. *Critique of Pure Reason.* Translated by Norman Kemp
Smith. New York: St. Martin's Press, 1965.

EM   Henry, Michel. *Essence of Manifestation.* Translated by Girard
Etzkorn. The Hague: Martinus Nijhoff, 1973.

GS   Kant, Immanuel. *Kants gesammelte Schriften.* Edited by the Preussischen Akademie der Wissenschaften zu Berlin. 29 vols. Berlin:
Walter de Gruyter, 1902–83.

Introduction

The monster in Mary Shelley's *Frankenstein* is monstrous because it is an act of blasphemy. By attempting to generate life by means of profane technological innovation, by attempting to generate spirit through profane matter, Frankenstein taints and perverts the sacred. Technology's attempt to reach the metaphysical level of the god-like is surely hubris. According to Walter Benjamin, Baudelaire, like many others, found the precursor to photography, daguerrotypy, to be "profoundly unnerving and terrifying" for much the same reason.[1] While we all may be playfully amused by stories of so-called primitives worrying that picture-taking would steal their souls, Baudelaire should make us think twice about the sense in which this is true. Benjamin's analysis of the aura of the work of art suggests that art is a living product that arises out of a dense, collective social web and that its meaning and function

---

1. Cited in Rutsky, *High Techne,* 37.

are bound to that context. Conversely, what Baudelaire found so frightening was that the communal context that was the very producer of meaning—as a kind of collective imagination, or what we might call spirit—was cut off from its artistic product. The technology of reproductive media, whether as daguerrotypy or photography, would usurp this role as producer. Andreas Huyssen invokes the rather apt term "technological imagination" as a way to grasp the manner in which the technology of reproductive media has both insinuated itself into and, in part, usurped the role of the imagination. He writes, "The invasion of the very fabric of the art object by technology and what one may loosely call the *technological imagination* can best be grasped in artistic practices such as collage, assemblage, montage and photomontage; it finds its ultimate fulfillment in photography and film, art forms which cannot only be reproduced, but are in fact designed for mechanical reproducibility."[2] Huyssen's point, like that of Baudelaire, Benjamin, and others, is that technology's insertion into the material practice of art is one in which the spiritual vocation of the human creative capacity of the imagination becomes appropriated and shaped—at least to some extent—by material technological practices. To return to Frankenstein and his monster, we might say that Shelley, like Benjamin and Baudlaire, feared that technology's appropriation of the spiritual vocation of the creative imagination would indeed produce the monstrous: art as the living dead. And while photography has not stolen our souls in any direct sense, the mediation of nearly every aspect of our lives from industrial mass production to the impact of the technology of reproductive media on language itself was a concern not only for Benjamin but also for other inheritors of the German idealist tradition, like Nietzsche and Heidegger.

Nietzsche's self-described role as a cultural physician makes clear the intimacy between theory and physiology. And while his own embodied need for such doctoring is well known in view of his own illnesses and physiological deficiencies—like impending blindness—what is less well known is the way in which Nietzsche courted technology both as a support and supplement for theory (for a hemorrhaging metaphysics) and as a buttress against his own debilitating physiological conditions. If, as Havelock Ellis argues, philosophy itself began with Plato's embrace of the technology of writing, Nietzsche's work in overturning Platonism was in part connected to his own technologically dependent style that would further divorce the spirit from the letter, the

2. Huyssen, *After the Great Divide*, 9.

Platonic ideas from the act of production, and ultimately metaphysics from technology. The technology of the typewriter made philosophy itself possible for Nietzsche by allowing him to continue to work. He writes, in reference to his Malling-Hansen typewriter: "Finally, when my eyes prevent me from learning anything—and I have almost reached that point! I will still be able to craft verse."[3] Yes, thanks to the typewriter. Yet this technology did not merely allow Nietzsche to continue working, it shaped his very philosophy by helping produce his "telegram style."[4] As he himself acknowledges, as a kind of confession, "Our writing tools are also working on our thoughts."[5] If Nietzsche seems to embrace a philosophy in which the disabled spirit of metaphysics requires technological support, Heidegger's wariness of technology hints of nostalgia. He famously writes: "It was not German Idealism that collapsed; rather, the age was no longer strong enough to stand up to the greatness, breadth and originality of that spiritual world."[6] One problem with nostalgia, of course is that it tends to yearn for a world that never was. As we will see, German idealism did not simply collapse: the relation between its metaphysical discourse of spirit and technology is far more complex than the language of mere collapse would indicate. And a detailed analysis of Fichte's work makes this unquestionably clear.

Around five in the evening, on September 14, 1813, in a salon at number 36 Franz Strasse in Berlin, Fichte witnessed firsthand a demonstration of the phenomenon of magnetic psychology or mesmerism.[7] Magnetic psychology was a material account of spirit that theorized that the whole world was imbued with a quasi-material invisible fluid that could be redirected to cure the sick and reestablish associations of collective social harmony. To the extent that many, like Fichte, defined spirit in terms of the imagination, its material technique was understood as a technology of the imagination. It took numerous forms, from a kind of learned material technique, much like contemporary hypnotism, to actual technological inventions, like the *baquet,*

3. Cited in Kittler, *Gramophone, Film, Typewriter,* 206.
4. Ibid., 200.
5. Ibid., 210.
6. Heidegger, *Introduction to Metaphysics,* 45–46.
7. Fichte's own experience of magnetic psychology is attested to in his unpublished "Diary of Animal Magnetism" ("Tagebuch über den animalischen Magentismus"). As I argue throughout this book, Fichte's in-depth knowledge of magnetic psychology, as well as his brief references to its terminology in earlier works, such as the *Grundlage des Naturrechts* (1796), suggests that not only did he have a lifelong concern with the phenomenon, but—as I will argue—it would play a crucial role in his formulation and attempted formulation of his lifelong project of the Wissenschaftslehre.

a large wooden tub with ropes, rods, and water, meant to link—materially—multiple individuals both to each other and back to the harmony of this quasi-material universal fluid.

Fichte's account of this event in his "Tagebuch über den animalischen Magnetismus" could easily be construed as an insignificant biographical detail if he himself had not seen magnetic psychology as key to warding off the rising tides of materialism, to preserving the dignity of spirit, and to preventing the total collapse of German idealism. The fact that Fichte would ultimately recognize that magnetic psychology could save neither his project of the Wissenschaftslehre nor German idealism's broader claims of spirit is far less relevant and interesting than his prescient attempt to ameliorate the infirmities of the idealist claims of spirit by technological means. And his failure is our failure to the extent—as Heidegger, Benjamin, and Nietzsche make clear—that it is a task with which we are still engaged today.

1

An Introduction to the Crisis of Spirit

Technology and the Fichtean Imagination

Reproductive Media and the Crisis of Intersubjective Recognition

J. G. Fichte is widely regarded as the first to articulate intersubjective recognition (*Anerkennung*) as a theory of right, and thus offer the notion of intersubjectivity as a significant social principle.[1] While Fichte's version of a theory of right is complex, the core principles of the notion are that one's rights end where someone else's nose begins. While such a gross simplification does little justice to Fichte's thought, in schematic form this account of individual right nevertheless sketches the essential intent of any reciprocal theory of right: it demands intersubjective recognition. It demands one individual recognize another individual as a free autonomous being. Here we have the

---

1. While the notion of intersubjectivity is implicit in Kantian ethics, particularly in the notion of a kingdom of ends, Fichte's transcendental deduction of the other was the first explicit expression of the idea of reciprocal recognition in the constitution of right.

essence of Fichte's conception of intersubjective right: "Each limits his freedom through the possibility of the freedom of the other" (*FW* 3:120–21).

Fichte's intent in outlining a theory of right was to deduce the conditions
of the possibility of freedom or free action. In the social sphere, he reasoned,
the freedom to act depends upon a reciprocal recognition of rights. One
rational being must recognize and take into consideration the rights of
another rational being as an autonomous subject.[2] Making space for the
rights of another through the free limitation of one's own action makes freedom possible. But what does it mean to recognize the rights of another
(another rational being)? What are the limits of recognition? And—perhaps
most important—how is recognition achieved? Such questions of recognition, which probe the very limits of Fichte's transcendental method, are further complicated by the technologies of reproductive media.

Consequently, with the gradual acceleration of the growth of technology
since the turn of the twentieth century, the very parameters of subjectivity
and its sphere of right, understood as a sphere of free action, have come to
be questioned. For instance, given the power of video, photography, and
television, whether one owns his or her own image has become an urgent
ethical question.[3] Thus, with such reproductive media it is not clear whether
a subject's sphere of right (conceived analogically as an extension of the idea
of physical bodily integrity), and the description of that sphere, can be limited in terms of an immediate sphere of free bodily action.

Technology has had an immense impact on Fichte's account of both self-
consciousness and intersubjective recognition. After all, what is unique about
Fichte's account of right is that its deduction of the possibility of freedom is
also a deduction of the possibility of self-consciousness. In other words, this
making space in the recognition of right gives space to the self-active ego,
and in the act of recognition makes self-consciousness possible. Thus, intersubjective right describes both the preservation of the self-active transcendental subject and its constitution in self-reflection. Consequently, one might
say that it is the form of social interaction, in this instance, the reciprocal
recognition of a community of free beings, which determines the form of
consciousness.

---

2. In my view Fichte's central concern with and understanding of right was determined by his
larger interest in the sociopolitical question of freedom. "Right," then, for Fichte, is primarily a right
to free action and freedom.

3. There has been increasing interest in the ethics of the image. See, for instance, Gross, Katz,
and Ruby, *Image Ethics*.

Although Fichte would initially describe the ego in terms of act, with the media image subjectivity can no longer be bound exclusively to a sphere of action and a system of rights that would secure it.[4] In an era increasingly dominated by technologically reproduced images, a theory and definition of right anchored in the free, self-limiting, and self-reflecting action of the transcendental subject seems in danger of being overwhelmed by the needs and demands of the *actual* forms of social interaction.

Fichte himself, as attested to in the "Tagebuch über den animalischen Magnetismus" (1813), encountered just such a crisis in the early part of the nineteenth century. What he encountered was a critical disjunction between his own theoretical account of intersubjectively constituted right and the actual lived form of subjectivity manifest in social experience. What this disjunction signals in Fichte, I contend, is a crisis of the transcendental imagination.[5] To understand this, however, we have to consider in detail the link between intersubjectivity and the imagination.

From Benjamin to Fichte: Technology and the Imagination

As I suggested with the example of the photographic image, the breakdown of intersubjectivity stems from the technological power of reproduction. Now one must imagine that the technological power to produce images threatens, or at least complicates, antecedent powers of imaging. Specifically, I am thinking of that distinctly modern power of imaging called the imagination, *Einbildungskraft*. While certainly there has been a long history of representation in Western art, or in the iconography of Western religious traditions, such images are mere products. And as products, they have remained subordinate to that productive/creative source called the faculty of the human imagination.[6] The imagination of the author or artist remains the original source

---

4. In Fichte's words: "All relation of right is determined through the sentence: each limits his freedom through the possibility of the freedom of the other" (*FW* 3:120–21). Consequently, right, and by extension freedom, is determined through a self-conscious, reciprocal, self-limitation.

5. While I will detail Fichte's understanding of transcendental philosophy shortly, for the moment it is sufficient to understand "transcendental" as a description of the power of the mind—or in this instance, the power of the imagination—to constitute experience.

6. The essence of the modern imagination is identified most readily in the notion of the creative and productive power of the individual. While the premodern artistic imagination was understood through the metaphor of the mirror in which an essentially anonymous individual (or group) functioned as a mimetic reflection of the divine. By contrast the imagination of the modern artist is understood through the metaphor of the lamp. Here the artist is his or her own productive source of illumination. See, for instance, Kearney's *Wake of the Imagination*.

of that work. The turning point comes, however, through a reversal; a reversal whereby the imagination is no longer the productive source of its world, but rather a subordinate and passive consumer of images external to it. Such a transformation, I contend, takes place in the work of Fichte. Yet just how the transformation of the imagination affects the dynamic of intersubjective recognition and right is perhaps nowhere clearer than in the work of the twentieth-century cultural critic Walter Benjamin.

In his essay "The Work of Art in the Age of Mechanical Reproduction," Benjamin addresses the problem of intersubjective recognition in the context of his larger discussion of the decline of the aura in the age of the mass.[7] The decline of intersubjective recognition is coincident with the decline of the aura. And although Benjamin does not use the language of "imagination" in this essay, his account of the decline of the aura describes the very same reversal of the imagination I just described.

The aura, for Benjamin, designates that radiant sphere of meaning which the singularly unique art object carries with it as part of its essential being. This meaning implicitly references that unique sociocultural horizon out of which a given "art" object is produced. The art object is the product of a determinate individual imagination—which is itself the product of a particular cultural horizon. The aura then designates a complex, dynamic, living web of meaning, whose sense-giving power stands in a relation with this singular created object as whole to part. The aura describes the object's meaning in terms of its living affiliation with its origin or creative source. To speak of the decline of the aura is to suggest that the object's meaning will come to stand apart from its origin or creative source.

And what Benjamin is suggesting is that there is an immediate link between the decline of the aura and the obsolescence of intersubjectivity as a social theory. But how? For him, the aura implies a recognition. Thus, in the same way a recognition of right confers a special status upon a particular being, as an autonomous being worthy of a space of free action, so too does the notion of the aura offer or confer a unique field of meaning to the object in question. At least this seems to be what Benjamin is suggesting. As he explains in "The Work of Art in the Age of Mechanical Reproduction," with technology, the power of imaging is unhinged from the structure of recognition in which "the glance" is returned. He writes: "The camera takes a person's picture without returning his glance. The glance, however, expects

7. See Benjamin, *Illuminations*.

inherently to find a response wherever it gives itself. Wherever this expectation finds such a response, . . . it experiences aura to the fullest" (*Gesammelte Schriften* 1/2:646). Thus, Benjamin concludes that with the decline of the aura in the age of the mass what is no longer returnable is the look, the defining moment of intersubjective recognition. Conversely, only where there is recognition does one experience the aura. Thus, with the decline of the aura in the age of the mass, the experience of recognition is forever altered. With the camera the glance is not returned; our expectation of recognition is not fulfilled.[8] Such a phenomenon, however, is not limited to the inhuman "seeing" of the camera and other technical equipment. For Benjamin, the camera is but an extreme instance of his more general point: with the decline of the aura, even common cultural objects lose their horizon of meaning, and as such, do not return the glance or affirm one's own sociocultural identity.[9]

In part, Benjamin intends to show us that with the decline of the aura in the age of mechanical reproduction the power of imaging is no longer simply the work of the imagination; it is also the effect of technological reproduction.[10] With mechanical reproduction the object no longer exhibits an aura; it is displaced from that determinate cultural horizon that could align that singularly produced art object with the creative, productive imagination of a unique author (or authors).

In brief, the object no longer exhibits an aura because the work of the imagination is externalized through material technique. This is a form of alienation. With mechanical reproduction what is eroded is that reciprocal relation of intimate recognition between those cultural objects an individual uses and those objects he or she creates. The situation, however, is still more extreme. It is not simply that the creative imagination is no longer the source of the objects an individual uses; but rather, foreign "use-objects" come to determine and give shape to the very the workings of the imagination.

Such a claim is grounded in one's everyday experience of television or

8. Sam Weber emphasizes in his work the importance of Benjamin's account of the decline of the aura for understanding contemporary issues in media, although he does not consider it in terms of an issue of intersubjectivity. See Weber, *Mass Mediasaurus.*

9. One might imagine, for instance, the intense identity-affirming experience of receiving, and wearing, a handmade sweater from one's grandmother, while living abroad in a foreign culture. What makes that initial experience abroad so alienating is that those foreign cultural objects stand outside one's own sphere of meaning and thus will not return the look. By contrast, the handmade sweater returns the look, a look of recognition, that affirms one's identity on the most intimate of levels.

10. Although Benjamin does not use the language of the crisis of spirit, his exploration of the crisis of the transcendental imagination, through the problem of intersubjectivity in the age of technology and technological reproduction, can be understood as a meditation on the crisis of spirit.

film. For instance, as a film spectator, one need not organize one's own narrative experience. The horizon, perspective, and even intentionality of that experience are already predetermined by the form and content of the technology of film.[11] My central point for our discussion of Fichte is that, as a consequence of such technologies, the transcendental subject and its imaginative faculty could no longer always secure a determinate locus, and the sphere of right, which was the end of intersubjective recognition, would be challenged as the fundamental designation of the social. In other words, in the same way that the aura of the object depends upon a determinate, irreplaceable locus for its meaning, so too is recognition contingent upon the self's sense of place as determined by that power of imaging (*Einbildungs-kraft*) anchored in the transcendental subject: it designates that purportedly inalienable place which defines the sphere of right of both oneself and the other.

What intersubjectively constituted right is confronted with is an entirely new power and structure of representation. The productive imagination, a mediating power of the self-active, self-constituting subject, was primarily a metaphysical description of the social dynamic of intersubjective recognition and the ascension to consciousness outlined in Fichte's *Grundlage des Natur-rechts*. In this sense, the crisis of intersubjectively constituted right, brought on by technological reproduction, signals a deeper metaphysical crisis of the imagination and its power of representation. Such a crisis of the imagination, however—as is clear from our references to technology and the problem of intersubjective right—is an issue with far-reaching consequences. As we will see, such a crisis had significant ramifications for Fichte's understanding of self-conscious reflection, imagination, intersubjectivity, and spirit.

I have just suggested that the social exchange or medium known as inter-subjectivity is intimately bound with a unique form of the imagination so that a transformation of the imagination, through its externalization in repro-ductive media, led to a decline in the predominance of that form of social intercourse known as intersubjective recognition. Yet building on the insights of Michael Theunissen, Jürgen Habermas, and others, I suggest that the decline of intersubjectivity was perhaps, in part, already latent in idealism's dual characterization of spirit as both intersubjective and as that immediate

---

11. Baudry emphasizes this point. See Baudry, "Ideological Effects." While certainly one might argue that a predetermined narrative structure was already present in the written text of the novel, the technology of writing, as we will see, is merely an earlier instance of the very same point or claim I am making with reference to film.

communicative transparency commonly associated with *Geist*.[12] What reproductive media brings to the table is the hope of a possible completion of metaphysics in technology through the material incarnation of spirit's dream of a pure communicative transparency. At least that is what the phenomenon of magnetic rapport seems to represent to Fichte: the very proof of his metaphysics of imaging. While I will return to explicitly address the transformation of spirit through the possible completion of metaphysics by technology, this decline begins as a crisis of spirit, at the impasse in the conflict between materialism and idealism. And it is only by traveling through this impasse that one can achieve the proper perspective to return to confront the crisis of intersubjectivity and the transcendental imagination in detail.

The Historical Horizon: Spirit in the Age of Materialism

It was not the material technique of magnetic rapport alone that complicated Fichte's account of the social and the transcendental imagination. This phenomenon was but a symptom of a larger crisis: scientific and technological materialism was beginning to threaten the very tenets of the idealistic conception of spirit. In its most basic form the clash between idealism and materialism was a confrontation between two different explanatory paradigms. Idealism's vocabulary of spirit was part of the discourse of an immaterial power and influence of soul and mind. By contrast, materialism best explained the causal influence between material bodies. The problem was that each respective explanatory paradigm was rather limited. Spirit could not readily explain causal interaction, just as materialism could not convincingly explain immaterial phenomena, such as mind. What was needed was a more universal and coherent explanatory paradigm. What was needed was a middle term between immaterial and material worlds, spirit and matter.

The middle term between spirit and matter has been a perennial problem for philosophy. For instance, the modern period has seen a series of unsatisfactory explanations of the soul's interaction with the body, from Descartes's "pineal gland" to Freud's "psychic representative." But how can one describe the point of intersection between a material body and an immaterial mind or soul? How can substances from two radically different orders interact?

---

12. I will discuss spirit in much greater detail shortly—both in terms of Theunissen's contribution, and in terms of the conflict inherent in the historical understanding of *Geist* as both *nous* and *pneuma*.

It seemed there was no convincing explanation. Although materialism was excellent at explaining the causal interaction among physical bodies, it was much more challenged in explaining consciousness and the immaterial nature of mind. Since its paradigm was material, consciousness would be explained—quite unsatisfactorily—as a series of chemical exchanges within the brain. By contrast, idealism's spiritual paradigm could explain the immaterial world of mind with ease, but it had much more difficulty in explaining the interaction between material and immaterial realms. Fichte, and much of nineteenth-century speculative philosophy, recognized some concessions to materialist science were necessary—if not inevitable. Nevertheless, the Cartesian mind/body dualism held certain advantages for both materialists and idealists. This dualism, for instance, offered a powerful mechanistic explanation of the world, thereby pleasing materialists, but at the same time it refused to cower before materialism's potentially extreme reductivist and eliminativist tendencies. It accomplished this balancing act by offering the rather obscure notion of mental substance, which allowed Descartes to establish the certainty of the *ego cogito,* while—perhaps inadvertently—protecting unexplainable aspects of mind from a reductive materialist explanation. The problem of dualism, however (even with the contemporary notion of "intentionalist property dualism," which suggests the uniqueness of intentionality in thought processes), is that it still cannot explain how material and immaterial worlds communicate.

The problem for some was that Descartes's affirmation of two different orders of substance (material and mental) described the mind as a wholly intellectual entity. This description, however, tended to obscure the more spiritual articulation of mind which it hoped to replace. The theological language of soul and spirit was used to describe many things, including the mental events of hoping, believing, anticipating, and so on, which we attribute to mind. Scientific materialism, however, would attempt to distance itself from the theological explanation in two key ways. While its primary intent was to explain what was formerly a function of the soul in the physical or material terms of nerve and brain function, it would also help rearticulate the discourse of the soul in terms of the mind within its own framework in order to explain those purely immaterial aspects of the mind which eluded materialist explanation. Of course, the language of mind was also present in theological tracts—such as those of Augustine and Aquinas. The difference, however, is whether mind is articulated as an aspect of materialist science or an aspect of the soul. Note the emphasis in each description. In the first, the material body is the central mode of explanation in which the term "mind" stands as a peripheral catch-all phrase.[13] In the second, by contrast, the discourse of the social is the key explanatory mode, where mind functions as a determinate aspect of the soul.

Our historical legacy of the language of soul and mind harks back to its source in the earliest Greek description of spirit as *nous* and *pneuma*. *Pneuma* was understood as a divine, naturalistic, animating life force (with some similarity to the Christian notion of soul).[14] Greek philosophy, however, for the most part eviscerated spirit by rendering it as *nous,* as the *logos* of reason and intellect.[15] While even here, and throughout the Middle Ages, intellect is very much a spiritual phenomenon, the intellectualization of spirit had nevertheless begun. Consequently, it is not such a long step from Augustine's conception of soul as a spiritual substance to Descartes's designation of mind as mental substance. One might even ask, more broadly, whether technology is a continuing step in the intellectualization of spirit from *pneuma* and *nous.*

The mind/body split is the legacy of *nous,* the intellectualization of spirit. One approach to the mind/body problem, then, would be to sidestep the very issue of dualism by returning to the notion of spirit as *pneuma. Pneuma,* as an animating principle, designated a form of spirit that was also material. For instance, for the Gnostics, *pneuma* was at once physical and spiritual: "Their spirit was fluid matter."[16]

Subtle Matter

The romantics—inspired by German mysticism—called for a return to the earlier cosmic-naturalistic conception of spirit. One dominant conception of a universal, spiritual, animating life force was that proffered by magnetic psychology. Magnetic psychology, founded by Anton Mesmer (1735–1815), claimed that all bodily and psychological disease was a function of an individ-

13. I am speaking more generally here, without specific reference to Descartes. Yet even with Descartes, where the *cogito* is central, one can see (curiously enough), that the mind is defined—by negation—in direct relation to material substance. Matter is extended substance, whereas mind is nonextended substance. It is whatever is *not* matter.

14. While soul and *pneuma* are both animating principles, a central difference between these two is that soul is an individual animating principle which functions in a transcendent religious framework, whereas *pneuma* is an anonymous life force which stands within an immanent, pantheistic-like religious framework.

15. See Berdyaev, *Spirit and Reality.*

16. It is much like magnetic psychology's notion of subtle matter. Ibid., 23.

ual's magnetic flow going out of balance with the larger natural flows of the earth and universe. While aspects of Mesmer's claims, as well as his personality, may have been conducive to charges of charlatanism, the broader field of dynamic psychiatry had much to offer. The field was in a state of constant development and refinement, and some of Mesmer's followers (like Armand de Puysègur and J. P. F. Deleuze) would develop the important psychological notion of an "affective rapport."[17]

The affective, communicative rapport between two or more individuals was made possible through a communication across spiritual and material worlds. What allowed this interaction across what I characterized as the divide of mind/body dualism was the notion of "subtle matter." The notion of subtle matter, as third term between spirit and matter was, of course, itself not new. In fact, it could even be understood, from a genealogical perspective, as a development of the articulation of spirit as *pneuma*. It can also be understood as part of romanticism's attempt to rearticulate aspects of spirit as *pneuma*.

What subtle matter designated was a middle term, a term that could mediate between material and immaterial worlds by embodying characteristics of each. Subtle matter was subtle enough to convey spiritual properties, yet could still exhibit the causal properties of matter. For instance, how the initial thought of moving my arm could actually produce physical movement was explainable by means of the mediating power of this third term. In the face of the surmounting difficulties of explaining the interaction of mind and body, recourse to a spiritual substance that moved effortlessly across the registers of spirit and matter was certainly a compelling idea.

With the nineteenth-century discourse of subtle matter, however, the mind/body problem was readily extended to the body politic. The question of communication between spiritual and material registers was now not only an issue of the mind/body relation; it was also a question of the other. In the social sphere the mind/body problem was interpreted as a question of intersubjective communication. After all, how could spiritual communication among individuals take place without the mediation of the material body? For Fichte the phenomenon of subtle matter had much to offer his own explanation and understanding of intersubjective recognition.

17. This rapport is considered a precursor to hypnotic rapport and the psychoanalytic notion of transference, which was a key element in the history of dynamic psychiatry, particularly in understanding the dynamic of intersubjective communication. The history of dynamic psychiatry refers to the particular historical genealogy that led to modern-day psychoanalysis. See, for instance, Henri Ellenberger's now-classic work, *The Discovery of the Unconscious.*

As I suggested earlier, Fichte would turn to the rival social theory of magnetic rapport and its language of subtle matter as an explanatory support for his own transcendental account of intersubjective right. Not surprisingly, Fichte's account of intersubjectivity exhibits the weakness of idealism more generally: it cannot explain communicative recognition—an influence that would proceed across the registers of material and immaterial worlds. It appeared that magnetic psychology's language of subtle matter was perhaps best suited to handle such a task.

Now what is crucial for the issue before us is that this form of influence designated by the medium of subtle matter would have far-reaching consequences for the idealist conception of the imagination. As recounted in both Fichte's own firsthand observations and extensive study of the phenomenon of magnetic rapport, as attested to in his unpublished work, "Tagebuch über den animalischen Magnetismus," the imagination, like spirit, would become embroiled in the confrontation between idealism and materialism, spirit and matter. As I already noted, Xavier Léon showed that Fichte's later description of the metaphysics of the imagination (whereby the subject stands as a conduit for the imaging of the absolute) was greatly influenced by the experience of rapport. As described in the "Tagebuch über den animalischen Magnetismus," what Fichte sought was a "Physicirung des Idealismus," physical proof for his idealist claims (TaM 70). For Fichte, magnetic rapport and subtle matter held out the hope for a mediation of spirit and matter. While Fichte ultimately recognized that rapport cannot truly stand as a material "proof" for idealism, and in particular his own later metaphysics of the image, he nevertheless experiences rapport as a material technique of the imagination, a technique whose reproductive power might seriously threaten his later metaphysics and the security of philosophical truth.

Spirit for Fichte

Like the Aristotelian notion of Being that is simultaneously everything and nothing, "spirit" in German idealism resists definition. Even in Hegel, where spirit is the central unifying concept, its meaning remains obscure. For instance, Robert R. Williams has suggested that "the more cynical may harbor the suspicion that not even Hegel knew what he really meant" by the term "spirit."[18] Fichte's notion of "spirit" seems both more and less obscure

18. Williams, "Hegel's Concept of Geist."

than Hegel's. It is more obscure insofar as spirit is not the central philosophical concept for Fichte as it was for Hegel. In fact, he rarely mentions it. It is less obscure, however, to the extent it is not all things for all purposes. More broadly speaking, while the idealist notion of spirit is as much a concept of reason as a dark pneumatological entity, Fichte's use of the notion of spirit is typically limited to the description of an immaterial communication.[19] Fichte references spirit when speaking about the problem of communication between two or more spiritual beings in the material world. Thus, he invokes spirit in two central instances: (a) in the possibility of spirit's mediation by the letter, and (b) in describing the very essence of the productive and representational structure of the imagination (as a power of communication).

In fact, Fichte goes so far as to align spirit and the imagination, defining spirit, like the imagination, as the power to project images. Fichte writes: "Is there anyone who cannot see . . . that the ability to project images is exactly what we have already described as spirit?" (*EPW* 204). And Fichte in the *Wissenschaftslehre* of 1794 writes, "The whole enterprise of the human spirit issues from the imagination" (*SK* 250). Spirit appears to be part and parcel of the imagination. In this respect, at least according to Fichte's definition, technology's impact on the productive imagination is equally a transformation of the spirit.

For Fichte the imagination is the source of spontaneity: it is the source of life. Its productive and representational powers make both consciousness and self-consciousness possible. As such it is the ground of all intersubjective communication. Here, the soul is read psychologically as a faculty, as the faculty of the imagination. Its role as both source of the spontaneity of the ego and as a communicative power tends to mirror the dialectical dependency of the I-we relation explicit in intersubjective recognition. As such, the very idea of spirit seems to be both singular, as soul, and communal, in the more common idealist understanding of spirit as communal spirit and universal category. In fact, as Hegel scholarship is well aware, we have two competing models of spirit before us. Jürgen Habermas, Michael Theunissen, and Ludwig Siep have all recognized that—at least in Hegel's work—intersubjectivity threatens the idealist model of spirit.[20] What this means is that the later Hegel constructed his model of absolute spirit as a development of transcendental

19. Alan Olson describes the importance of the pneumatological articulation of spirit for German idealism in his work, *Hegel and the Spirit: Philosophy as Pneumatology.*

20. See Habermas, "Arbeit und Interaktion"; Siep, *Anerkennung;* and Theunissen, "Die verdrängte Intersubjektivität."

subjectivity, as an absolute subject that was opposed to his earlier articulation of spirit as a communicative medium. Theunissen explains: "Hegel is never able to clarify satisfactorily the relation between the *Geist* which functions as a medium, and the *Geist* which is the self-consciousness which comprehends this intersubjective mediation."[21]

My intent with this excursus into Hegel is to highlight how the very same conflict is present in Fichte's work. Intersubjectivity implies a form of dialectical mediation that, in principle, can never achieve spirit's ideal of pure communicative transparency or immediacy without its own self-overcoming. In this sense spirit and technology must dismiss intersubjectivity in the name of the same ideal: pure communicative transparency.

The very notion of spirit embodies an ideal of pure communication. Spirit to spirit communication, however, is an ideal that is without instance in the finite material world. Here, if any communication is possible, spirit must somehow be mediated by matter. Still, communicative technology's intent is to fulfill spirit's dream of communicative transparency in material terms. And it is in this sense that we must ask to what extent technology can be understood to (aspire to) fulfill the dream of metaphysics. Fichte too encountered this question in his own repeated confrontation with the impasse between idealism and materialism, spirit and technology.

Spirit's encounter with materialism provoked a crisis. And this crisis of spirit, I argue, was essentially a crisis of communication. The problem at hand was much like the question of the interaction between mind and body. After all, how could human communication, and thus the spiritual life of the community, exist in a materialist science's description of a collection of causally interacting bodies? As we will soon see, while Fichte himself resorted to the language of mechanics, as with the "check" or *Anstoß*, his discussion of the summons and recognition in the *Grundlage des Naturrechts* ultimately demanded a more "spiritual" or at least psychological discourse.[22]

Since the work of spirit is indistinguishable from that of the imagination, spirit's encounter with materialism can readily be understood as an encounter between materialism and the imagination. Such an encounter is perhaps nowhere clearer than in Fichte's later discussion of the paradigm of subtle matter and magnetic psychology in the "Tagebuch über den animalischen Magnetismus." While I will develop the idea of subtle matter still further,

---

21. Theunissen, "Die verdrängte Intersubjektivität."
22. I discuss this terminology in detail in Chapter 4.

what is significant is that this term represents *a material technique for the formation and direction of the imagination. What Fichte encounters is an early technology of the imagination.*[23]

I will conclude this work by arguing that magnetic rapport stands as the culmination of what I call the double displacement of the transcendental imagination. In briefest possible terms, I see the double displacement of the imagination as follows. In the first moment, in the transition from Fichte's first to last period, the transcendental subject was displaced when Fichte reversed the priority of subjectivity and Being. Fichte's later metaphysics does not characterize the imagination as a faculty or locate it in the transcendental subject. In the *Wissenschaftslehre* of 1813, the imagination is presented as arising in Being's process of self-manifestation, and Being itself as coming to appearance through the transparent I (*das durchsichtige Ich*). This first displacement of the imagination, however, was soon threatened by a second one. Here the metaphysical account of the externalization of the imagination is threatened by the material technique of magnetic rapport. I contend that if the metaphysics of the imagination is reproducible through material technique (and stands as a material proof, incidentally, which Fichte's very conception of philosophy required), then the ontology of the imagination is threatened by a technology of manifestation. To review, then, while it is generally acknowledged that the transition from a discourse of an ego defined as act to one defined in terms of image signals a clear deposing of the transcendental subject and displacement of the transcendental imagination, Fichte scholarship has not recognized the severe consequences such a displacement provoked at the dawn of technological materialism.

Nevertheless, Heidegger, of course, was well aware of the danger and consequences of such "placeless-ness." For instance, one of Heidegger's central worries in the "Age of the World Picture" was that without the ground of the transcendental subject the imagination's *productive* power of phenomenal manifestation was in danger of being aligned with the technological power of *reproduction*. In more Heideggerian language, technological enframing (*Gestell*) infringed upon the free appearance of Being.[24] I argue that this was

---

23. I am not alone in characterizing the phenomenon of magnetic rapport as a technique of the imagination. Some of Anton Mesmer's own contemporaries described it in these terms. For instance, see Charles d'Eslon's essay of 1784, "Observations on the Two Reports of the Commissioners Named by the King to Investigate Animal Magnetism."

24. Of course, as Heidegger's entire critique of ontology insists, Being can only come to appearance through beings (*Seiende*). Further, as J. Dreschler, W. Janke, and others have noted, Heidegger's phenomenological ontology has much in common with Fichte's own later metaphysics of imaging. See Dreschler, *Fichtes Lehre vom Bild,* and Janke, *Fichte: Sein und Reflexion.*

also one of Fichte's central worries, that he too perceived a crisis of the imagination stemming fundamentally from certain techniques and technologies of manifestation that threatened to eclipse ontology's power of revelation. In the face of this threat, the human imagination could no longer be conceived—without suspicion—as a divine medium that was the very source of a productive human creativity. Rather, since the power of imaging might now be merely the effect of human reproductive technique, the question of philosophical truth now becomes threatened by the veil of imitation. As we will see, in this sense, Fichte attempted to secure philosophical truth and the sociopolitical order much as Plato had in his diatribe against imitation in *The Republic*.[25]

This threat to truth strikes at the heart of Fichte's fear of the social and political consequences of a materialization of the imagination. Thus Fichte, like Heidegger, I argue, resisted the materialization of the imagination, not as a denial of his own sociohistorical conditions, but as an as ethical imperative. To hold such an imperative, however, is not always to call for the rebirth of the idealist narrative of spirit. Rather, such a position, in its embrace of an "outside," might stand merely as a form of resistance. In "The Question Concerning Technology" Heidegger himself would claim that "the essence of technology must harbor in itself the growth of the saving power."[26] His point is that spirit and technology, and thus perhaps idealism and materialism, are not unequivocally opposed. For instance, aspects of technology might hold within them a redemptive capacity for a new articulation of spirit.

Social Interaction and the Transcendental Imagination

The crisis technological materialism poses for Fichte's conception of intersubjective right has not been adequately recognized as a crisis of the imagination. Nevertheless, I will show how the power of imaging embodied in the technology of reproductive media threatened Fichte's notion of the imagination as a faculty, *die Einbildungskraft*, and the forms of social interaction, recognition, and self-conscious subjectivity that follow from it.

While many scholars—like J. Dreschler, W. Janke, X. Tillette, and R. Lauth—readily acknowledge a development in Fichte's view of the imagina-

25. See, for instance, books 7 and 10 of *The Republic*.
26. Heidegger, *The Question Concerning Technology*, 28.

tion from the first to the last period of the Wissenschaftslehre as a transition from subjectivity to Being, few make explicit Fichte's encounter with materialism, which, I will argue, heralds a second and significant displacement of the imagination.[27] Xavier Léon, however, stands as an exception. In *Fichte et son temps*, Léon makes two points that are crucial for the argument at hand. First, he points out that it was the threat of materialism in the voice of Karl Solger (1780–1819) that first led Fichte to search for a physical proof for idealism. Second, he suggests that the very idea of a transparent I (*ein durchsichtiges Ich*)—which is central to Fichte's late metaphysics of the image—was inspired by Fichte's own experience of the transparency of the subject in the phenomenon of magnetic rapport.[28] Léon's two claims—first, that Fichte's later metaphysics was influenced by, and in dialogue with, the rise of materialism, and second, that Fichte sought a mediation of the claims of idealism and materialism in magnetic rapport and magnetic psychology's notion of subtle matter—mark an important starting point for any understanding of the role of technology in Fichte's account of the social.[29] These observations show that Fichte was indeed aware of the threat of materialism and that he sought a means for some sort of mediation beyond the strict confines of idealist and materialist paradigms.

Nevertheless, Léon fails to follow up on the implications of such provocative observations. For instance, he suggests that Fichte's very notion of a transparent I (*ein durchsichtiges Ich*) was gleaned from the phenomenon of rapport. And although Léon infers that magnetic rapport was—as an inspiration for Fichte's later metaphysics—continuous with Fichte's later theory of the imagination, I make the stronger claim that the paradigm of magnetic rapport represents a second and distinct displacement of the imagination.

In other words, if—through the transition from subjectivity to Being—Fichte's first displacement of the imagination can be understood as a deposing of the transcendental imagination, then the second displacement of the imagination (as witnessed in Fichte's engagement with magnetic rapport) arises through the externalization of the imagination, through the constitution of the power of imaging through material technique. The imagination

27. Although scholarship on the later Fichte is limited primarily to studies on the Continent, this view could be understood as the canonical or accepted interpretation.

28. Magnetic rapport, understood broadly, describes an affective, communicative influence between individuals. The details of such a theory, as it originated with Anton Mesmer, will be outlined shortly.

29. Léon, *Fichte et son temps*, 2:283.

is displaced for a second time when Being's power of manifestation is usurped through technological means. If in the first displacement the representational power of the imagination was displaced and externalized so that the subject was no longer the source of the imagination but a mere conduit of it, the second displacement of the imagination shows that at least in regard to the power of imaging, technology has the power to imitate what was formerly a metaphysical ontology. This claim of a double displacement of the transcendental imagination forms the foundation of my main argument: *intersubjectively constituted right is no longer possible because the representational power of the imagination was externalized and usurped by technological means.* In order to work out such claims and to understand in detail this double displacement, we first need to understand what the transcendental imagination is for Fichte.

According to Fichte, the Wissenschaftslehre is through and through a project of transcendental philosophy (*GA* 3/3:331–33). Now Kant had understood transcendental philosophy to be a knowledge of our a priori mode of understanding objects. Thus, he was concerned "not so much with objects as with the mode of our knowledge of objects insofar as this mode of our knowledge is to be possible a priori" (*CPR* B25). Fichte attempted to carry out the spirit of Kant's project. Thus, Fichte defined "the very essence of transcendental philosophy" as an activity that "does not directly engage in representing, but rather represents the activity of representing" (*GA* 2/3:325–26). One important difference between Kant's and Fichte's respective accounts of the transcendental is that while Kant emphasized the product, the knowledge of a priori knowledge, Fichte was perhaps more concerned with just *how* the transcendental project was to be carried out. Fichte was concerned with the possibility of transcendental philosophy as a process.[30]

Now to speak of transcendental philosophy seems to imply a transcendental subject. In other words, although we can speak of the transcendental imagination in Fichte as what constitutes experience a priori, this constitution is nevertheless the experience of an experiencing subject. After all, it would be meaningless to speak of a form of knowledge without reference to a knower. If transcendental philosophy is *knowledge of* the a priori forms of knowledge, then it would seem (necessarily) to require a form of reflective self-consciousness. When Kant, in the *Critique of Pure Reason*, explains the transcendental unity of apperception by stating that "it must be possible for

30. Wayne Martin emphasizes this point in his work. See Martin, *Idealism and Objectivity*, 64–65.

the 'I think' to accompany all my representations" (B 131–32), he is affirming the necessity of self-conscious reflection for the transcendental account of knowledge. Consequently, for Kant, a representation without the accompanying "I think" is not knowledge, but mere intuition.

From an interpretative standpoint, Fichte's approach to apperception is a much more difficult and disputed issue. While for Kant all consciousness was representational consciousness, Fichte saw things differently. Fichte offered a pre-representational mode of consciousness called "intellectual intuition." In the "Second Introduction to the Wissenschaftslehre," Fichte explains that intellectual intuition is the "intuition of a sheer activity" which provides not self-knowledge, but self-certainty (ipseity) (*IWL* 48).

For Fichte the ego need not be representational because the understanding and its representational power are no longer the source of transcendental spontaneity. This is now the work of the imagination. Thus what one could call apperception is now achieved through a pure and spontaneous self-positing. The I, in its immediacy, is simply posited—along with the Not-I. The I and the Not-I form two unquestionable first principles, which one either affirms or rejects (*IWL* 14). As such, the imagination is not the ontological source of the I (or the Not-I); rather, as the source of transcendental spontaneity, it stands as an active intermediary between these terms, thereby giving shape to experience (*SK* 194). In more distinctively Fichtean language, the dialectical oscillation (*schweben*) of the imagination is key to the transcendental ego to the extent it makes possible the transition from the prereflective immediacy of "intellectual intuition" to the full-blown reflection of self-consciousness.

From Intersubjectivity to Rapport: Imagination and the Critique of Self-Conscious Reflection

If we are to understand the crisis reproductive media poses for intersubjective right—and on a greater scale, the crisis technological materialism poses for spirit—then we must return to this basic question: what does intersubjectively constituted right have to do with the imagination? To begin with one needs to recall that a recognition (*Anerkennung*) of the call of the other (*Aufforderung*), which provides a free limitation of one's own action, is a reflective process: it both delimits the parameters of one's subjectivity (in terms of right) and makes possible the inception of self-consciousness. As such, intersubjectivity would seem to be a condition of self-conscious reflection. Thus, with the development of reproductive media and the consequent breakdown of intersubjectively constituted right, one must wonder whether the structure of self-conscious reflection is itself threatened. Or, conversely, from the side of self-conscious reflection, one must wonder what sort of social theory a critique of self-conscious reflection would imply.

The crisis of the imagination brought on by the technology of reproductive media is first witnessed by Fichte at that moment in which the imagination is manipulated by material technique. Although Fichte's most elaborate discussion of that material technique occurs in 1813 in his discussion of magnetic rapport in the "Tagebuch über den animalischen Magnetismus," his references to the vocabulary of magnetic psychology occur as early as the *Grundlage des Naturrechts* (1796). In the *Grundlage des Naturrechts*, at those very moments when his own explanation of the possibility of intersubjective right begins to falter, he supports his transcendental account through reference to magnetic psychology's notion of "subtle matter."

In the nineteenth century, subtle matter stood as a key mediating term in the attempt to explain the communicative interaction between material and immaterial worlds. Subtle matter stood as an explanatory key in the description of a communication between mind and body, and between the spiritual and material worlds. In Fichte's instance—at least in the *Grundlage des Naturrechts*—what he needed to explain was how, in recognition (*Anerkennung*), an immaterial intellectual or spiritual awareness could have a material or causal-like influence upon another rational being (as distinct from any other sort of being) (*FW* 4:301, 3:80–81).[31]

Fichte's attempt to buttress his transcendental account of the other through reference to an affective rapport seems to anticipate his deposing of the transcendental subject in his emphasis on Being in the last period and his growing interest in the importance of magnetic psychology for his own work (as attested to in the "Tagebuch"). Fichte's repeated references to subtle matter in the *Grundlage des Naturrechts* mark the limit of a transcendental explanation.[32] Consequently, I will argue that such references (a) expose the shortcomings of his intersubjective account of right, (b) highlight the difficulties of a strictly self-conscious, reflective model of subjectivity, and (c) ultimately bear themselves out as a crisis of the imagination.

31. It was in many ways simply a unique form of the problem of action at a distance in which the entire science of magnetism was steeped.

32. See *GNR*, sections s4–s7.

Subtle matter's more primordial explanation of the possibility of social interaction through an affective rapport seems to stand as an extension—*as the social facet*—of Dieter Henrich's more epistemologically driven critique of self-conscious reflection. In other words, Henrich's emphasis on Fichte's prereflective, self-affective self seems to dovetail with the affective paradigm of social interaction that magnetic rapport represented.

Henrich, in his highly influential article "Fichte's Original Insight," critiques the theory of philosophical reflection as an explanation of self-conscious subjectivity by pointing out the inherent paradox of reflection. Philosophical reflection's explanation of self-conscious subjectivity by means of self-reflection, he contends, necessarily falls prey to paradox.[33] Now, as I understand it, the two central paradoxes Henrich points out are much like the classic philosophical problem referred to as Meno's Paradox. In the *Meno,* Plato points out to his interlocutors that the fundamental, but—as he ultimately shows—merely apparent paradox of knowledge is as follows.[34] If one had complete knowledge there would be neither the desire nor the need to seek it. By contrast, if one were wholly ignorant one would have neither the sense nor the desire to even begin on the path to knowledge. One could not even begin a first step of inquiry. Henrich sees the paradox of reflection in much the same way. On the one hand, if one is able to reflect back, observe, and recognize oneself, then one is already self-conscious and such an act would be a meaningless exercise. It assumes what it sets out to explain— namely, the self of self-consciousness. By contrast, if one has no intentional reference or criteria by which to recognize the self that is to be observed, then self-consciousness would be an impossibility. One could never come to recognize the self, even having stumbled upon it.

This question of self-conscious reflection returns us once again to the issue of apperception. Plato argues that Meno's Paradox is not real, but merely apparent, because one is never really ever in a position of complete knowledge or complete ignorance. It seems to me—despite Henrich's articulation—that Fichte's answer is much the same. The relation of knowledge to action is such that we are never without at least a mute sense of the self. By contrast, the ideal of self-reflection, pure transparency, like Platonic knowledge, remains (for Fichte also) essentially an ethical or regulative ideal. In this sense

---

33. Henrich, "Fichte's Original Insight," 15–53.
34. See Plato, *Collected Dialogues.*

reflection is not the paradox it initially appears to be, provided, however, one offers—as Fichte does—an account of an ascent toward self-consciousness beginning in self-feeling. Yet despite the fact this paradox is more apparent than real, Henrich's article nevertheless forced Fichte scholarship to take seriously the status of the prereflective "I" and the role of self-feeling in the account of self-consciousness, freedom, and free action.

If Henrich's critique of reflection has at least highlighted the significance of the prereflective, self-feeling self in Fichte's work, then we need to address how such an articulation bears on the crisis of intersubjective recognition. In other words, in what sense can an affirmation of the self in terms of a prereflective, self-feeling self still be read as an implicit critique of the transcendental model of intersubjective recognition? If the self-feeling self or the self of intellectual intuition stands as a support, as the anterior moment, in the development of self-consciousness, how might one understand Fichte's own references to the vocabulary of an affective rapport in his account of intersubjective recognition in the *Grundlage des Naturrechts*? What I am suggesting is that Henrich's critique of reflection and affirmation of a self-affective self— despite its shortcomings—is quite suggestive for my own articulation of the crisis of intersubjective right—which Fichte ultimately experiences as a crisis of the imagination. In other words, although Henrich's critique has its difficulties, his emphasis on the importance of a self-affective, prereflective self for Fichte's account of self-consciousness seems to imply that the description of intersubjective right might also demand some sort of prereflective, affective ground.[35] Not surprisingly, Fichte's own references to subtle matter in the *Grundlage des Naturrechts* seems to bear this out.

Fichte's references in the *Grundlage des Naturrechts* to affective rapport as a support for his own theory of intersubjective right suggest and even anticipate his own explicit recognition of the inadequacy of the transcendental model. In this sense, Fichte's "original insight," I argue, was his insight into the theoretical demands of this new form of social interaction. Still, it is no surprise that Henrich did not move in this direction. As is clear from the cool public reception received by Schelling's own account of the prerational basis of transcendental philosophy, transcendental idealism—strictly speaking—is not particularly friendly to theories of the unconscious; what this latter model

35. For some of difficulties scholars have had with Henrich's essay, see Neuhouser, *Fichte's Theory of Subjectivity,* and Pippen, *Hegel's Idealism.*

suggests is precisely an unconscious (*unbewußt*) or preconscious explanation of the social.[36]

Fichte's own reinterpretation of Kantian apperception led him to a prereflective, self-affective self in the spontaneous act of self-positing. Since such an action was merely self-determining and not truly self-creating (in the strict sense) (*FW* 1:126–130), as Schelling pointed out, even with this "self" the problem of its condition of possibility looms large.[37] Yet despite Fichte's repeated disputes with Schelling over the question of the ground of the transcendental ego, Fichte seems uniquely aware of the problem. For instance, in the "First Introduction to the Wissenschaftslehre," Fichte designates this unknowable ground—the very source of the I that stands beyond consciousness—as the fundamental point of investigation for his philosophical research. Fichte writes, quite paradoxically, that "philosophy must discover the ground of all experience," although "its object necessarily lies outside all experience" (*FW* 1:425; *SK* 8). The paradox, he is well aware, is that "the ground falls outside what it grounds" (*FW* 1:425; *SK* 8). The key to discovering this ground—by means of what it grounds—lies in the recognition that the ground and grounded "are opposed but linked" (*FW* 1:425; *SK* 8).[38]

For Fichte the intimacy of the ground/grounded relation, that is "opposed but linked," defines a limit condition, a condition that described his fundamentally opposed but mutually dependent propositions of the I and Not-I (*FW* 1:123–130). In brief, Fichte's definition of the ego as act delineates a dynamic of striving and limitation that makes possible the self-feeling of the original self.[39] The feeling of limitation constitutes an original self-feeling (*Selbstgefühl*), a feeling of the self anterior to self-reflection's grasp of itself as an object. Fichte clarifies this distinction between self-consciousness and consciousness as follows: "This immediate consciousness is the conscious subject in every *act* of consciousness, *but it is not the subject of which we are conscious* {Nothing that we can be conscious of is immediate consciousness

---

36. Like the prereflective self's relation to self-consciousness, the affective account of the social does not intend to replace the account of intersubjective right, but merely shows it to be a significant and a more primordial form of social interaction.

37. Schelling, *Sämmtliche Werke*, 1:80.

38. The relation between the ground, as the condition of the possibly of experience, and conscious experience itself is made clear in Freud's own polar topography (in the conscious/unconscious dyad). One can only read the unconscious, from this side of consciousness, through the slips and ruptures in the patient's speech.

39. One of many instances in which Fichte claims the ego is activity is when he notes "the intellect, for idealism, is an act, and absolutely nothing more." He also claims "the self is an original doing" (*FW* 1:440–41; *SK* 21).

itself . . .}" (*WLnm* 128; my italics). Following Kant, Fichte aligned this original preconscious, unconscious, or nonconscious (*unbewußt*) self with the idea of a "noumenal self," a self beyond the appearance of a determinate consciousness (*WLnm* 408). Yet what is our relation to this noumenal ego, if, as ground, it itself can never be known?[40] As Fichte suggests, because the ground can only be approached by means of the grounded—by invoking the game of the limit—it makes sense to embrace self-awareness in its most primitive form, as feeling.[41] Yet despite this limit condition, Fichte continually attempts to catch himself by the tail. He tries to lay bare the ego's condition of possibility by means of the notion of the self-seeing eye. Fichte establishes the relation between the immediate consciousness of feeling and self-consciousness proper by means of the analogy between the eye itself and its act of seeing (*WLnm* 129). The dream of the self-seeing eye, however, is a vision that would haunt Fichte until the end of his life.

We must not forget that the ground of intersubjectivity and self-reflection originates in the transcendental imagination. In order to better grasp the role of the imagination in self-reflection and intersubjectivity we must delineate the relation between the *Grundlage des Naturrechts* and the *Wissenschaftslehre*. Like our awareness of the other that defines that self-limiting activity called "the sphere of right," in a more metaphysical register it is the self-limitation of the *Anstoß*, of the Not-I, which delimits the ego and establishes the field of subjectivity through the oscillating activity of the imagination. In fact, in the *Wissenschaftslehre* of 1794, Fichte defines the imagination as an "absolute activity that determines reciprocity" (*SK* 150) in its oscillation (*Schweben*) between self-striving and self-limitation. Like the self-limitation that arises from a recognition of the other, the *Anstoß* is a check, an impetus toward

---

40. Since for Fichte "the self is originally a doing," one would have an intimation or a feeling of the original ego in the act itself. In fact, Fichte (as we noted) defines the ego as act. And action for Fichte is a dialectic of striving and resistance. Thus, it is in this dialectic of action that a self-affective, or prereflective, primitive self-awareness first arises. It is through the encounter of resistance or limitation that the striving self first arrives at a primitive self-awareness or feeling. While the resistance to the striving self merely "is," one's own gradual awareness of this limitation is what Fichte calls "feeling." Limitation and self-feeling, however, are only the first steps on the path to self-consciousness (steps, incidentally, whose full exposition falls well beyond the scope of this limited investigation).

41. The path to self-consciousness proceeds as follows: limitation, feeling, intellectual intuition, and self-consciousness. Since Fichte's notion of intellectual intuition is quite different from Kant's, and is further prone to misunderstanding, particularly in the context of a discussion of the unconscious, I would like to differentiate it from feeling. With feeling, one is aware of limitation. Intellectual intuition, however, designates a metaperspective: it is the feeling of feeling as one's own (*WLnm* 192).

self-determination. In this sense, in the *Grundlage des Naturrechts* the delimitation of social subjectivity as a sphere of right is closely tied to the explanation of the transcendental ego by means of the imagination in the *Wissenschaftslehre*. In other words, the *Wissenschaftslehre*'s metaphysical account of the transcendental subject in terms of the constituting power of the dialectical imagination (in its oscillation between the I and Not-I) is mirrored in the social sphere in the account of intersubjective recognition.

The theoretically conceived oscillation of the dialectical imagination has a practical aspect whose description, like the notion of *Anstoß*, is profoundly physical in character. Here, *schweben* is interpreted as a dialectic between striving and limitation. As such, the imagination takes on the character of a psychological and motivational power of action. By "raising feelings to consciousness" (*EPW* 194) the productive imagination is a source of spontaneity that establishes its own directed limitation through the self-active limiting effect of the *Anstoß*—which makes possible both representational consciousness and (through the power of reflection) transcendental subjectivity. My point in discussing the constituting power of the imagination for the ego is to emphasize that the crisis of the imagination brought on by the techniques of technological reproduction will have a real impact on the formation of the subject and the delimitation of its sphere of right.

Still, the precise nature of the relation of the account of self-conscious activity found in the *Grundlage des Naturrechts* to that found in the *Wissenschaftslehre* is in some dispute. For instance, both Ludwig Siep and Wilhelm Weischedel see the *Wissenschaftslehre* as a foundation of which the *Grundlage des Naturrechts* is but an instance or practical example.[42] By contrast, Alexis Philonenko, Peter Baumann, and Reinhard Lauth see the *Grundlage des Naturrechts*'s explanation of intersubjectivity as either already present in the *Wissenschaftslehre* (Philonenko), or as a quasi-independent project implicitly acknowledged and assumed by the *Wissenschaftslehre* (Baumann and Lauth).[43] This debate, which I will detail again shortly, remains situated primarily in the first period of the Wissenschaftslehre. Taking the last period of the Wissenschaftslehre into consideration, and in particular the influence of the social dynamic of magnetic rapport and subtle matter in Fichte's later

42. See Siep, *Anerkennung*, and Weischedel, *Der frühe Fichte*.
43. See Baumann, *Fichtes Ursprungliche System;* Lauth, "Le problème de l'interpersonalité chez J. G. Fichte"; and Philonenko, *La liberté humaine dans le philosophie de Fichte*. Robert Williams gives an excellent overview of literature on this problem in his book *Recognition: Fichte and Hegel on the Other*.

description of imagination and subjectivity, one wonders whether (in the last period) if it is not the form of social interaction itself that dictates the more metaphysical speculations of the last *Wissenschaftslehre,* rather than the reverse. In fact Xavier Léon will insist that, in the last period of the Wissenschaftslehre, it was Fichte's own firsthand engagement with magnetic rapport that led to the displacement of the transcendental imagination and the embrace of a passive subject that was a mere conduit of Being's process of imaging. In short, following Léon's insight, I suggest that in Fichte's later theory of the imagination, the relation between metaphysics and social theory was not in dispute: it was the form of the social that ultimately dictated the shape, direction, and culmination of Fichte's own metaphysical ruminations.[44]

44. Léon, *Fichte et son temps,* 2:283.

2

Technology and Truth

Representation and the Problem
of the Third Term

Introduction: Technology, Mimesis, and Truth

In the end, Fichte recognized that the success of his philosophical system, and transcendental idealism more generally, would require a mediating third term between spirit and matter. What is at stake in this mediating third term is truth itself: and his search for this term must therefore be developed within the context of his philosophical and historical search for truth. I propose that Fichte's own difficulty in coming to terms with subtle matter as a possible proof for his metaphysics can be understood on the model of Plato's own difficulties with mimesis in *The Republic*.

The issue of mimesis highlights philosophical truth's continuing battle with the Homeric oral tradition out of which it rose. And it is this conflict between philosophy as a discipline and its origin in the "affective" aspects of the oral tradition that I believe bears directly on Fichte's understanding of the imagination, both in terms of how technology—and the very technology of print culture—can be understood to complete philosophy, and in terms of how Fichte's own later conception of truth, as a metaphysics of imaging, is conceived on the order of affect (*FW* 5:498). And although mimesis appears to be in conflict with the philosophical notion of truth as *adequatio*, it is precisely what I call its aesthetic truth, truth as revelation, that makes it most fitting in accounting for Fichte's later metaphysics of the imagination. Further (as I already suggested), it is Fichte's discussion of his own metaphysics of revelation in terms of affect (*FW* 5:498), and his ultimate desire for a physical proof for such metaphysics (TaM 70), which requires us to confront the notion of subtle matter and magnetic rapport as a problem of technology that finds its root in the very issue of Platonic mimesis.

Mimesis and Representation

Technology's reproductive power of imaging poses a threat to the imagination. This threat, however, is not merely social and political. Because the very nature of intersubjective right, and thus the medium of the social, would be transformed by technology's usurpation of the transcendental imagination, this crisis also occurs as a threat to philosophical truth. Now the imagination, as a power to produce images, is a power of representation.[1] With the displacement of the imagination, the communicational medium of the truth—which I will speak of in broadest possible terms as "representation"—undergoes a major transformation. The power of representation becomes unhinged from the transcendental subject. As a consequence, the very nature of representation, and its relation to truth, is profoundly altered. In brief, if communication, the communicational medium of truth in the form of representation, is what makes possible a social and epistemological community, then technology's usurpation of the imagination's power of representation would radically transform both community and the communicational medium of truth.

The meaning of the phrase "the communicational medium of truth," however, will remain elusive until we better understand the nature of the medium itself. This medium, I will argue, can be understood fundamentally

---

1. In the body of this chapter I will better define the sense in which I use this term "representation."

as an issue of representation. Now the historical problem of representation—perhaps conveyed most incisively in Plato's discourse on mimesis in *The Republic*—exemplifies the problems inherent in any middle term.

As we shall see, how one conceives representational truth is essentially a question of loyalty. Like the middle term(s) of magnetic psychology, which would be asked to mediate between spirit and matter—and thus take part in each—so too in the classical philosophical problem of representation will the middle term be asked to stand between the immaterial world of the idea and the material realm of appearance. Admittedly, this middle term, because it cannot actually "become" truth, is typically accused of conspiring against truth and allying itself with imitation.

The quest to present the immaterial and infinite aspects of metaphysical truth in and through a finite medium has posed innumerable challenges to philosophy and the discourse of representation. The search for a third term in the disciplinary registers of both philosophy and magnetic psychology was a demand for an *appropriate* third term. One that could fulfill the impossible demands of, on the one hand, conveying the truth without being the truth, and on the other, of conveying the infinite and immaterial aspects of spirit, without being spirit.

Like the middle terms of magnetic psychology, the challenge placed upon philosophical representation—to convey the immaterial and infinite aspects of truth from this side of material appearance—would demand its own undoing. The theoretical challenge of representation then would be to describe or delimit a "liminal" representation; one that could stand on both sides of the material/immaterial divide in the process of its own undoing. As Kant recognized with the sublime, representation can only achieve certain ends as a consequence of having exceeded its own limits. While such a phenomenon is clearly visible in the register of rapport, in the patient's transparency to the will of the magnetist, it is still clearer in Fichte's later metaphysics. Here the absolute comes to manifestation through the transparency of the subject (*das durchsichtige Ich*). The absolute comes to appearance in and through the subject's own dissolution. The subject itself comes to stand as liminal representation. Such a phenomenon, however, is not without precedent. These "liminal representations" can be understood within the context of the greater philosophical crisis of representation from Plato's notion of mimesis, to Kant's account of the sublime, to Fichte's own attempt to delimit a prereflective sense of the self in terms of *Darstellung*. In brief, then, the struggle to articulate the self-creative self at the very limit of representation is an early moment in the crisis of representation and truth which finds its source in the imagination.

The quest to mediate truth, which is the very task of the imagination, can readily (and perhaps too hastily) be interpreted as a technique or technology. For Fichte, this mediative and imitative role is precisely the danger of technology. Yet insofar as the imagination is grounded in the creative subject and its liminal representations embody aspirations for truth, such a conflation ought not to occur. Nevertheless, the imagination's own difficulty in mediating truth by means of a proper liminal representation allows the medium, technique, or technology to take precedence and efface the original task of the creative imagination. Thus, in this chapter, the medium will be delineated as a technique or problem of technology through an analysis of Fichte's understanding of representation—particularly in those historically contingent modes of representation called rhetoric and writing.

Plato and Mimesis

Plato outlined a complex analysis of representation through a development of the notion of mimesis. While an elaborate explanation will be undertaken shortly, for the moment it is sufficient to recognize that the paradox of mimesis was that it was a form of imitation that, as a unique medium, harbored within it the possibility of truth. Like good and bad forms of rhetoric, mimesis could also be said to have a use-value. In other words, what makes mimesis so important to our discussion of the Fichtean image as a medium is that mimesis—like Fichte's discussion of the appearance (*Erscheinung*) of the image (*Bild*)—is a dissimulation that may also be a revealing of aesthetic truth.[2] Of course, like the illusory nature of appearance, it may also be merely the mask of a lie. In this respect mimesis, the mimetic essence of the image, is not only a dangerous medium of truth—for finite beings, it stands as perhaps the only medium of truth. This is the problem of the technologization of the imagination. If the power to produce images is no longer the work of spirit—as a revelation of the absolute—its revelatory capacity may be less a work of spirit than one of mere dissimulation.

Plato's own articulation of truth's relation to technology through the

---

2. As we will discuss in more detail shortly, what is unique about mimetic truth is that it stands beyond the framework of an *adequatio*. And it is this circumvention of a correspondence theory of truth that frees mimesis from the discourse of imitation.

medium of rhetoric and writing has an immediate bearing on Fichte's encounter with the crisis of the materialization of the imagination, insofar as this crisis can be traced back to Fichte's own first grappling with technology's relation to metaphysical truth in the discourse of spirit's relation to the letter.[3] As Plato makes clear, this conflict is essentially an issue of mimesis. After detailing the issue of mimesis in Plato, and the broader stakes of such concerns within the history of philosophy (particularly as described by T. W. Adorno), I will show how mimesis is ultimately translated into the discourse of idealism in the vocabulary of representation (specifically, as *Darstellung*). As we will see, *Darstellung*, as a form of representation, is a process of sensibilization (*Versinnlichen*) that might stand as a third term between spirit and matter.[4] Thus, it is Fichte's first attempt to delimit a liminal form of representation. As such, *Darstellung* is a term that will be key in Fichte's own desire to articulate a self-active self anterior to a reflective model of self-consciousness. I conclude this chapter with an anticipation of Fichte's own exploration of representation and the medium of truth in terms of the impact of technology, the technology of writing, upon spirit.

Mimesis, Technology, and Truth

Any attempt at a genealogy of mass media communications and its technology of mediation must recognize that the question of the medium over and above the message has been a fundamental political issue ever since Plato, in the *Phaedrus*, attempted to expel rhetoric from the purity of philosophical *logos*. Here, Plato disparagingly described rhetoric as "the art of conducting souls by words" (266e–267d).[5] Technology's relation to philosophy is made clear when Plato accuses the Sophists of instrumentalizing the *logos* by means of *techné*.[6] Nevertheless, the irony was not lost on Cicero, who pointed out

3. Even for the early Fichte, the letter represented a technology of mediation for the conveyance of spirit.

4. In this chapter *Darstellung* will be contrasted with *Vorstellung* and *Nachahmung*.

5. Plato, *Collected Dialogues*.

6. Châtelet, *Platon*, 65. *Techné*, which designated a general "art" in Homeric times, took its first step toward our present understanding of technology as an abstract, mediated technique through the following distinction. As Francois Châtlet points out, *techné* is distinguished from *epistemé* in a unique political context. The distinction arises when the philosopher argues that the Sophist is merely instrumentalizing the *logos* through rhetoric in order to seize power—without regard for truth or knowledge. This etymology is the basis for my own understanding and description of rhetoric and rapport in terms of technology.

that Plato's victory over the Sophists was attributed to the rhetorical superiority of his (Plato's) discourse. Plato, of course, like Aristotle, recognized the importance rhetoric held for truth and thus established a distinction between good and bad forms of rhetoric.[7] What Plato feared was the dangerous power rhetoric represented, and the possibility of this power falling into the wrong hands.

Rhetoric, as the *techné* of word and image, was understood to be an obscure and dangerous art, one that constituted a dangerous social bond insofar as its mediation of body and soul resulted from the completed circuit between word and nerve. Yet to move beyond the *episteme* of *logos* and to embrace the rhetorical affectivity of the word would be to reconceive the social bond at the frontier of ideal and material worlds—at that instant at which word has become flesh and nerve. After all, how else could bodies be moved by truthless words? Yet the form or figure of this medium that would constitute this aforementioned "circuit" is always a historically determined one. Across theories of rhetoric, ether, invisible fluids, hypnosis, fascism, and the electronic mass media of television and Internet, we have endlessly attempted to represent, articulate, substantiate, and give visibility to the form of an affective social bond. Yet the Faustian desire to give visibility to this unseen medium harbors within it, as Plato was well aware, a yearning for power. Whoever could unlock the secret of this social medium could control his or her neighbor, and by extension, the polis.

While philosophy has always been wary of the affective power of rhetoric, it has also—ironically—exhibited an uncanny attraction for the organic materiality (*Materialität*) of words. The psychophysical juncture of word and nerve—whether as rhetoric or as subtle matter—has long represented something of philosophy's secret holy grail.

Consequently, the middle term between matter and spirit has been a perennial problem for philosophy. For instance, in the modern period the soul's intersection with the body can be said to have produced a series of unsatisfactory solutions to the problem of the middle term from Descartes's "pineal gland" to Freud's "psychic representative." In the eighteenth and nineteenth centuries, although the quest for a "spiritual matter" was nothing new, what had changed was the ever expanding scope and intensity of the search for material imponderables. A veritable social mania arose out of the search by speculative physics for subtle matter(s). And the list of impondera-

---

7. See Aristotle's *Rhetoric* 2.1.

bles whose theoretical characteristics could qualify them as candidates for
subtle matter grew rapidly in this era to include ether, dynamic forces, magnetic fluids, and hypno-suggestive speech.[8] This broad genealogy of the
medium or middle term between spirit and matter through this theoretical
and historical cartography should make it clear that the political and epistemological impact of subtle matter (which, for Fichte, marks the imagination's
first transformational encounter with technology) is, in essence, not significantly different from the larger problem of representation that has plagued
philosophy since its inception. The unique problem of representation, and
the difficulty in delimiting its sense, is that, like the Platonic problem of
mimesis, it stands as both the truth and lie of imitation.

Mimesis: From Philosophical to Aesthetic Truth

The term "mimesis" resists definition. Like the work of art to which it is
often applied, mimesis is a form of imitation, but one whose capacity for
bringing things to appearance and visibility also suggests a power for truth.
Through the production of phenomena it *represents* appearance. The difficulty before us, which is, in effect, the problem of mimesis, is what one
means by the term "representation."

   In order to answer such a question we must examine the relation between
truth and representation. Yet it is difficult to delimit representation's relation
to truth because truth itself is not easy to define. In schematic form, one
quickly realizes that the philosophical problem of representation, as a legacy
of mimesis, offers two conflicting paradigms of truth, the *philosophical* and
the *aesthetic*. Philosophical truth operates as a correspondence theory in
which the truth or reality of an object is adequated to (that is, measured
against) another that stands as the standard of measure itself. One sees this
in the notion of the Platonic Idea. In this theory an object's reality is determined by its proximity to the ideal form (the Idea). In such a framework,
representation or mimesis is clearly a second-order, degraded moment of
truth known as imitation.

   By contrast, aesthetic truth recognizes the insufficiency of the model of
the *adequatio*. As a revelatory or aesthetic theory of truth, it shows that the
appearance generated by mimesis is not reducible to imitation, but like the

8. Cantor and Hodge, *Conceptions of Ether (1749–1900)*, 19–30.

work of art, is in fact a creative, self-referential entity that is the source of its own truth. Here representation does not seem to be a simple copying in which one could confirm the status of imitation as *doxa*, through an *adequatio*, a correspondence model of truth. Rather, mimesis designates an artistic and poetic production that is not a mere representation of reality. It is a creative production, a *poesis*. For instance, the Plato scholar Richard Patterson points out that if a painter were commissioned to paint the fiercest dog, that work would not be diminished by the fact such a dog does *not* actually exist.[9] What such an example suggests is the powerful self-referentiality of mimesis. Its truth is not determined through an adequated subject/object correlation, but by an aesthetic truth somehow inherent in its own productive imitative essence. This is the paradox of mimesis: it is an imitative form whose self-standing, self-referential, productive capacity to bring things to appearance may offer a glimpse of truth. Plato's difficulty in eradicating mimesis from *The Republic* is testament to his own position regarding the tension between these two competing forms of truth.

Plato developed philosophy and analytic thought within the horizon of an oral culture by banning mimesis. The paradox is that he recognized the value of the aesthetic truth of mimesis, all while attempting to distance himself from it. This complexity of and confusion surrounding his position is perhaps nowhere more apparent than in his discussion of the forms of representation known as writing and rhetoric. Before I analyze the role of these representational forms in the generation of oral and analytic consciousness in specific terms, I will briefly review the role of mimesis in the development of rational consciousness from a still wider historical perspective.

On the Relation of Aesthetic and Philosophical Truth

T. W. Adorno and Max Horkheimer offer a broader approach to the development from mimesis to rationality in *The Dialectic of Enlightenment*.[10] In this work they argue that with the transfiguration of mimesis into rationality and the concomitant subject/object split of epistemology, what is lost is one's fundamental relation to nature, to ourselves, and to one another in the social order. What breaks down, through what might be described as epistemological alienation, is the intimacy between an individual's inner and outer worlds.

9. Patterson, *Image and Reality in Plato's Metaphysics.*
10. Adorno and Horkheimer, *Dialectic of Enlightenment.*

Mimesis first appeared as a mimicry in which humans established an identification with the natural world. An affirmation of identification was necessary because there was already a break from nature. Nevertheless, this
identificatory relation indicated the achievement of both self-consciousness
and self-enjoyment.[11] Yet with the gradual development of this imitative,
identificatory capacity arose ever more determinate forms of intentionality
that would become an "organized control of *mimesis*," a mimesis on the cusp
of rationality.[12] Here as mimicry developed into an intentional structure,
nature was taken as an object distinct from that intentional subject. This
separation of subject and object, which is the mark of rationality, is demarcated most clearly as *res cogitans* and *res extensa* by Descartes. The mimetic
relation that binds inner and outer worlds, mind and body, and oneself to
others is shattered through the process of objectification. Yet as Adorno
emphasizes, mimesis is not necessarily opposed to, but rather founds cognition. He writes: "Cognition itself cannot be conceived without the supplement of *mimesis*, however it may be sublimated. Without *mimesis* the break
between subject and object would be absolute and cognition impossible."[13]
Mimesis then is not the antithesis of cognition. And Adorno is not invoking
some romantic dream in order to return to a past that perhaps never was.
Rather, Adorno recognizes that insofar as mimesis stands as a condition of
the possibility of cognition, the contributions of mimesis, particularly as a
remedy to a technocratic rationality, ought to be recognized in order to fully
embrace the powers of an enabling rationality.[14] Mimesis then is an essential
moment in the movement from subject to the other, or outer world, an
originary moment in the development of the bridge in the movement from
one's internal to external world(s). The danger, Adorno recognizes, is that in
epistemology the trend away from mimesis leads to a radical alienation of the
subject from its object, its world, and its human other.[15]

11. Ibid., 54–55.
12. Ibid.
13. See Adorno, *Against Epistemology*, 143.
14. A technocratic or instrumental rationality, according to Frankfurt school intellectuals,
defines a means-end rationality whose subject/object alienation is so extreme it borders—as Witt-
genstein has pointed out—on pathology. One particularly gruesome example, of course, would be
the rather efficient instrumental rationality the Nazis employed in the Holocaust. By contrast the
discourse of an enabling rationality suggests that reason itself is not corrupt, but rather, like the
Kantian critical limitation of the reason's penchant for abstraction and reduction, its good must be
kept in check.
15. And it is here that mimesis, as a problem of truth, appears simultaneously as a political issue.
As we will see, Fichte encountered much the same problem.

Mimesis is profoundly political in at least two senses. On the one hand, it is a threat to the rational political order, as Plato suggested; on the other, it designates that fundamental affective, aesthetic, and often dangerous bond that is also the essence of the political—as Adorno suggests, and Plato, perhaps, well knew. As the original identificatory bond between self and other, mimesis could not be eradicated from the polis because it was essential not only to pedagogy but to politics. It is a residual mimesis that on the one hand makes epistemology itself possible and, on the other, prevents the subject from being doomed to solipsism, by standing as the very ground of (political) subjectivity.

Fichte's account of intersubjective right attempts to explain social interaction and recognition in rational terms, but it is an account, I will argue, whose gradual unearthing of the ground of mimesis in intersubjectivity in the form of an affective rapport leads to a crisis of philosophical truth. Before we develop this political aspect of intersubjective right in detail, however, we must return to focus our attention on the issue of mimesis, truth, and technology. Unlike the epistemological demands of the *adequatio*, the truth of mimesis, I suggested, is a self-referential, creative force, much like Spinoza's *natura naturans*. Its model of truth, like its model of politics, is fundamentally aesthetic. The truth that emerges from the self-creative, self-productive work of art need not have any corresponding reality, and in this sense is describable as illusion. As Adorno explains: "Art moves towards truth. It is not directly identical with truth; rather truth is art's content. Because of the relation to truth art is cognitive. It knows truth insofar as it manifests itself in art. As cognition, however, art is not discursive. Nor is truth the reflex of an object."[16] Art stands as a medium for the manifestation of truth. The question of mimesis and truth has arisen in our discussion of the Fichtean imagination because it is the *status* of the images as a medium that is at issue. And with the eventual usurpation of the imagination's power of imaging by technology, what is at issue is the very source of philosophical truth and the fate of the creative imagination. It was Fichte's own difficulties with the explanation of intersubjective recognition that led him to magnetic rapport, or in effect, to a more explicit description of the social in terms of mimesis. Nevertheless, such an explanation is a paradigm from which he will ultimately attempt to extricate himself in order to save philosophical truth.[17] For

16. See Adorno, *Aesthetic Theory*, 394.
17. Once again we are in the midst of the conflict between philosophical and aesthetic truth.

the moment, we must begin to delimit the status of the Fichtean image within the context of the question of mimesis. The fundamental problem of the communication of truth for Fichte—whose problematic is radicalized with the creative imagination's encounter with technology—appears as a discussion of representation (*Darstellung*), particularly in the context of the discussion of the relation of the spirit to the letter. Before we turn to develop Fichte's own view, however, I would like to further set up the background to this investigation by continuing our analysis of mimesis in Plato's framework of writing and rhetoric.

Plato, Writing, and Truth

The judgments that Plato, and Fichte in turn, make regarding rhetoric must be understood in the context of the rise of the technology of writing and other technologically mediated forms of (spiritual) communication. The relation of rhetoric and writing can be condensed into two distinct views, which, as representational forms, reflect the two aforementioned aspects of mimesis. The oral tradition, which is aligned with the view of truth as aesthetic truth, conceives speech, like mimesis more generally, to be an ambivalent form of representation whose poetic value might, in a revelatory fashion, reveal truth. By contrast, Plato, in an effort to avoid the imitative dangers of mimesis, attempts to distinguish philosophical truth from mimesis. As is well known, he does so by attempting to banish the mimetic art of poetry from *The Republic*.

In book 3 of *The Republic*, Plato first suggests that mimesis operates merely as a form of imitation, a "likening of one man to another" (393c). His argument, however, first, that no one can function successfully as a pure *mimos*, imitating all things, and second, that even in a particular art, true knowledge is superior to its pretension, is a form of argument that not only upholds an analytic criterion of knowledge but also vitiates the very possibility of a pure mimos from the start (395b). This is the strategy for *The Republic*. Plato writes that in this utopia "there's no double man among us, nor a manifold one, since each man does one thing" (397e), "nor is it lawful for such a man to be born there" (398a). As is well known, the problem that persists for Plato is that while he can ban the artists and poet-performers from *The Republic*, he cannot eradicate the mimesis essential to education, because education, and in particular the education of the guardians, remains an essential element in *The Republic.*

Plato first insists that the *paideia* (or *Bildung*) of the guardians involve no imitation, but then concedes that "if they do imitate they must imitate what is appropriate" (395c), like courage and freedom. It is an issue once again of propriety and impropriety. Mimesis then does have a profound value in *The Republic,* but it must be kept in check. It is a powerful tool in the formation of the guardians, but it must serve the ends of an ordered society. The greatest dangers to order are embodied in the imitation of slaves, madmen, and women, representing—respectively—the loss of autonomous selfhood, instability, and the hysteria of the feminine.[18] Mimesis left unchecked would disrupt the economy of representation that makes possible the epistemology and ethics of the autonomous self-consciousness that founds the philosophical project.[19] In fact, philosophical truth would come to define itself, negatively, as a break from mimesis, just as the rigor of the ordering of the written *logos* would come to define itself by distancing itself from the *lexis,* the predominance of the technique of the medium (in the oral tradition).[20] Thus, with the dismissal of mimesis as the essence of the oral tradition there is a dynamic, mutual stabilization of both subject and object; there is a developmental and complementary reciprocity between the self-conscious, autonomous subject and the definition of truth aligned with the stabilization of the object in an eternal Platonic form. And it is in this sense that imitation stands as a threat to self-conscious, autonomous subjectivity.

It should be noted that Plato also disparages writing. In *The Republic,* Plato levels at least three charges against writing. Writing is said (a) to destroy memory, (b) to weaken the mind, and (c) as a manufactured project, outside the mind and beyond the protection of its author/creator, it remains essentially indefensible and unresponsive (275a). Still, what are we to make of the fact that, despite Plato's objection to writing, his dialogues are some of the first documents of Western civilization committed to writing? And further, what are we to make of a thinker whose attack on both rhetoric and writing is made possible by the very forms he denounces?

Plato's rhetorical stance on the question of writing exhibits a strategy that lulls the defensive posture of his opponent by seeming to offer numerous

18. See Lacoue-Labarthe, *Typography,* 129.
19. Ibid., 124.
20. Plato uses the common Greek word *lexis* to refer to what we might translate as the verbal medium, mode, or style.

concessions, concessions which allow Plato to lump writing and speech into a problematic that he himself is at liberty to frame. According to Plato, writing, like rhetoric, is a question of "propriety and impropriety" (274b). Thus, although Plato seems to have framed the problem of writing in terms of the problem of rhetorical speech, and thus, in effect to have capitulated to his opponent, his determination of what ultimately constitutes propriety and impropriety outlines a form and style of thinking that could only be the effect of a consciousness and pattern of thought *already formed by writing*. The definition of what counts as real speech is not the discourse of an oral culture, but a style of speaking whose consciousness has been already been permeated by the technology of writing.[21]

Writing is a technology that determines the form of consciousness, and it is this analytic thought which makes philosophy itself possible. Walter Ong's assertion in *Orality and Literacy* that "writing is a technology" that "initiated what print and computers only continue"[22] can be viewed as a logical extension and important deepening of Eric Havelock's landmark work, *Preface to Plato*.[23] In this work Havelock argues that Plato's entire epistemology could be viewed as a strategic rejection of oral culture and the social forms of consciousness it fostered. This rejection arose as a consequence of an embrace of an linear, typographic, and analytic consciousness that was a direct product of the technology of writing.

Havelock argues that writing's emergence from oral culture established an analytic form of thought that was the hallmark of philosophy. Further, he maintained that, for Plato, because oral culture is imitative in practice, and thus, in essence, antithetical to the stability and unity of truth, such imitative practices must be ordered and controlled by philosophy and, by extension, the power of the technology of writing. From such an analysis one could infer that philosophy, as a form of analytic consciousness, is itself a technology.

As a consequence of this technology, the meaning of the Greek word *psyche* underwent a major transformation with the spread of literacy in the fifth century B.C.E.[24] As Havelock explains, it is the emergence of literacy that first made possible the conception of the soul as "the ghost that thinks," and bestowed the notion of selfhood, and with it the possibility of moral responsibility and moral cognition.[25] Or, as Havelock states succinctly, "The

---

21. See Ong, *Orality and Literacy,* and Havelock, *Preface to Plato.*
22. Ong, *Orality and Literacy,* 82.
23. See Havelock, *Preface to Plato.*
24. Ibid., 197.
25. Ibid.

doctrine of the autonomous *psyche* is the counterpart of the rejection of the oral culture."[26] Since, as Havelock argues, writing is what makes possible "selfhood" and the "autonomous psyche," then we must recognize that the resolution of the debate between the technology of writing and truth, particularly in Fichte, will ultimately determine the very formal articulation of self-consciousness. In brief, the question of the technology of writing cannot be separated from the issue of self-consciousness.

The analyses of Ong and Havelock, which argue that writing is a technology that formed human consciousness as analytic self-consciousness, are approaches whose critical account of a philosophical truth generated by writing seem to lament the loss of aesthetic truth as mimesis. In this context, Plato's lumping together writing with oral culture, defined as rhetoric, suggests that writing and rhetoric are both representational technologies. Such a conclusion would allow Plato to concede Havelock's and Ong's criticisms, but without allowing them a revelatory sense of truth, a truth beyond the technological structuring of self-consciousness and truth.

As we will see in Fichte's own discussion of truth and writing, in his works on the spirit and the letter of philosophy, what is at issue in the question of the technology of writing is the very representational medium of truth. Plato's definition of philosophical truth as *adequatio* largely reduced both speech and writing to a form of analytic self-consciousness determined by writing.[27] As we will see in the next section, Fichte, by contrast, will attempt to preserve an aesthetic sense of truth, a mimesis freed from the stigma of imitation. Fichte's fear of the imitative aspect of writing and aspiration for the immediacy of the truth of spirit led him to embrace the breath of spirit, or in other words, that part of the oral tradition which philosophy, in its ascension though the written word, would prefer to leave behind. Such a conflict of spirit, couched in terms of *pneuma* and *nous,* should be familiar to us from Chapter 1. Nevertheless, although Fichte recognizes spirit as the "primal sublime" (*EPW* 195), and thus as something akin to *pneuma,* some representation, some representational technique or technology, is, for him, necessary.

If we recall writing's constitutive power for self-consciousness and the immediacy associated with orality, Fichte's critique of writing must be under-

26. Ibid., 200.
27. The extent to which Plato concedes the necessity of mimesis within the Republic suggests his own recognition of the value and necessity of aesthetic truth. The intricacies of such a debate, nevertheless, would take us well beyond the concerns of our present project.

stood as an extension of his critique of self-conscious reflection, while his limited embrace of rhetoric and the oral tradition must be aligned with his attempt to articulate the sense of a prereflective self. In other words, for Fichte, while the technology of writing stands as a form of representation in the strong, imitative sense, and is, as Havelock claimed, a counterpart to self-consciousness, the rhetoric of oral traditions, by contrast, exhibits the structure of a productive mimesis: it is a vital source whose proximity to truth, the truth of spirit, is recognized by Fichte in his attempt to delimit a prereflective self in the discourse of *Darstellung*.

From Mimesis to *Darstellung*

The classical Platonic problem of mimesis is transformed into the modern issue of representation. In eighteenth- and nineteenth-century Germany an indication that the complexity of the issue of *mimesis* and aesthetic representation had not waned is clear in the multiple terms for representation. Here representation is described as *Repräsentation, Vorstellung,* and *Darstellung.* Admittedly, the distinct meanings of these different terms for representation are difficult to parse in any definitive fashion since historically they often have been used quite loosely. In eighteenth- and nineteenth-century romanticism and idealism, this distinction among terms, nevertheless, did designate a distinction in meaning. *Repräsentation* was a "making present" in a material and visual sense. It was derived from the Latin *repraesentatio. Vorstellung* indicated an image within the subject's mind. And *Darstellung* was a process of *Versinnlichen,* a "making-sensible" of the concept. It was derived from the Latin *praesentatio.* To give a further indication of the complexity of terms, *Darstellung* initially implied a rejection of mimesis and imitation, while Herder and much of the romantic tradition rehabilitated the term precisely as a translation of Aristotle's definition of mimesis.[28] Of course, since our interest is in Fichte, we need not be concerned with the intricacies of an extensive and detailed etymology, but rather, we are fundamentally concerned with his usage of these terms. Although, as will be clear from Fichte's appropriation of Kant's aesthetic use of *Darstellung* in the Third Critique, Fichte can also be understood to use *Darstellung* largely as a translation of the classical definition of mimesis.

28. See Lacoue-Labarthe and Nancy, *Literary Absolute,* viii, and Seyhan, *Representation and Its Discontents,* 7.

The Kantian definition of *Darstellung* will be crucial to Fichte's own artic-ulation of representation. Kant derived his notion of *Darstellung* from classi-cal rhetoric—from the Latin *exhibitio,* and the Greek *hypotyposis.*[29] Through such a definition what is emphasized is what is visually present to the eye. Hence, as Kant explains in the *Critique of Pure Reason, Darstellung* designates the givenness of the object in intuition (B195). While *Vorstellung* describes an empty concept, a concept without intuition, *Darstellung* for Kant explains a sensible intuition that is the product of the unity of intuition and concept.

One aspect that will be crucial for our discussion of representation in Fichte is the status of the representation of the "I."[30] As Kant explains in the First Critique, the subject cannot present itself as it truly is (B152–53). He writes: "Here is the place to explain the paradox . . . namely how this inner sense presents to consciousness . . . ourselves only as we appear to ourselves, not as we are in ourselves" (*CPR* B152–53). Since Kant is unable to unify the subject as it appears with the subject as it is, the "transcendental unity of apperception" is untenable unless the problem of representation can some-how be freed from the stigma of imitation through the notion of *Darstellung.*

Darstellung grows in significance for Kant in the Third Critique. In fact, his expanded account of *Darstellung* in terms of aesthetic representation sug-gests the possibility of at least a partial solution to the problem of the presen-tation of the subject "as it is" (B84) through his discussion of the beautiful and the sensible in terms of *indirekte* (*CPR* B125; *CJ* s59) and *negative Dar-stellung* (*CPR* B97; *CJ* s27). The negative *Darstellung* of the sublime compels the subject to think the supersensible. Thus, if Kant's account of appercep-tion in the First Critique shows the limits of *Darstellung,* the Third Critique can be understood as a *Darstellung* of limits.[31] As we will see in the later Fichte's concept of the transparent I, it is the appearance (*Schein*) of represen-tation (*Darstellung*) at its limit, at its moment of failure, and in its *Darstellung* of that failure that—perhaps ironically—it is most capable of transmitting the truth of the absolute. We see this, of course, in Kant's account of the sublime.

As Fichte recognized the issue of the representation of the I in the context of his explanation of the original preconscious self, the *Selbstdarstellung* of the I is a striving that exhibits a self-referentiality much like that of a work of art.

---

29. See *CPR* B240, B255. Rudolph Gasché makes this point in "Some Reflections."
30. While I broach the issue of self-reflection here, I analyze it in much greater detail in Chapter 4.
31. Helfer makes this point in her work. See Helfer, *Retreat of Representation,* 47.

While Fichte ultimately offered a self-positing self, he first described the self-creative, self-referential self in terms of *Darstellung* in his *Aenesidemus* review. Now although the story of this review is philosophically and dramatically complex and implicates an entire cast of characters (among them Kant, Reinhold, Schulze, and Fichte), for our purposes we need only look at how this review led Fichte to articulate the centrality of *Darstellung* for his own work. I will nevertheless offer the most rudimentary sketch possible of this controversy.

Reinhold's critique of Kant's dual source of cognition (in sensibility and the understanding) led Schulze to critique Reinhold, while Fichte in turn would come to defend the spirit of Reinhold's work by recasting his notion of *Vorstellung* in terms of *Darstellung*. As Fichte recognized, Reinhold made the mistake of attempting to unite Kantian sensibility and understanding by assuming *Vorstellung* was the most elemental aspect of consciousness. His mistake was as follows: with the term *Vorstellung* he cast the mind as a static entity and insisted this most elemental aspect to be a fact (*Tatsache*) of consciousness. In response, and as a corrective, Fichte suggests that the mind is not a *Tatsache,* but actually an act, a *Tathandlung.*

In both his early drafts of the *Aenesidemus* review and his "Own Meditations on Elementary Philosophy," Fichte recasts the issue of representation in terms of *Darstellung.* Here *Darstellung* is a creative, self-productive act, which he describes in terms of the *Tathandlung.* Both *Darstellung* and the *Tathandlung* can be described, according to Fichte, as "an activity which presupposes no object, but itself produces it, and in which accordingly the acting [*Handlung*] immediately becomes the deed or fact [*That*]." While such a description anticipates Fichte's own account of the self-positing I of the *Wissenschaftslehre* of 1794, the self-creative aspect of this definition of *Darstellung* is also reminiscent of the productive, creative aspect of mimesis as a *poesis.*

Fichte describes the self-active, self-creative subject in terms of *Darstellung.* He explains: "The subject is (for itself) by virtue of its being—this happens through being active, the being active is the source of being, and is also its effect; this activity is called [*Darstellung*] positing itself as self in existence; and the power *Darstellungskraft*" (GA 2/3:89). By defining the I as an original activity, as *Tathandlung,* in terms of the self-creative act of *Darstellung,* Fichte is able to explain—and thus, has explained—the source which represents the empirical subject as *Vorstellung.* As Fichte puts it: "Everything which enters into empirical consciousness is *Vorstellung; Darstellung* never enters into empirical consciousness; but it alone constitutes pure consciousness" (*GA 2/ 3*:153–54). As Fichte clarifies in "The Spirit and the Letter of Philosophy," the task of the aesthetic drive, whose fundamental mode is that of *Darstellung*, is the "creation of the image in the soul," where by contrast, the practical drive "intervenes in the order of representations and sets up a possible external and extraneous objective for this *Nachbildung* in the external world" (*GA 1/6*:343).

Through this distinction between *Darstellung* and more imitative forms, like *Vorstellung* and *Nachbildung*, Fichte attempts to resolve the difficulty Kant had encountered in the "transcendental unity of apperception" between the self as it appears and the self as it is. And he will attempt to resolve this difficulty through the notion of *Darstellung*. By means of this term Fichte offered an original presentation of the self-creative self, anterior to self-reflection proper. Fichte overcomes the paradox of the representational self by discarding the definition of representation in terms of correspondence (*adequatio*) through an aesthetic account. With the aesthetic drive (*ästhetische Trieb*), the aesthetic object (in this instance, the "I") is given complete autonomy and freed from the referent.[32] Thus, Fichte explains that the aesthetic drive is wholly self-referential. He writes: "No prior representation of its object is possible because its object is itself only a representation" (*GA 1/ 3*:345).

In the *Wissenschaftslehre* of 1794 Fichte shifts from *darstellen* to *setzen*. Here the I no longer presents itself (*sich darstellen*), but now posits itself (*sich setzen*). It seems Fichte shifted vocabulary because he thought of *Darstellung* as fundamentally an artistic description and hoped to maintain a distinction between philosophical and aesthetic discourses.[33] In fact, in his "Own Meditations on Elementary Philosophy" one can see Fichte experimenting with the vocabulary of *Darstellung* and *Setzung*. He toys with the idea of various word constructions, like *Dar-setzung* and *Dar-legung*. In this discussion his difficulty is clear. He writes: "The analyzing philosopher *legt dar* [expounds]: the poet painter sculptor *stellt dar* (the products of his creations, his fantasy)" (*GA 2/3*:89).[34]

Such an explanation along with Fichte's eventual return to the rhetorical tradition of *Darstellung* in 1801 should make it clear that Fichte never really

---

32. See *Über Geist und Buchstab in der Philosophie* (*FW* 8).
33. Helfer emphasizes this point. Helfer, *Retreat of Representation*, 73.
34. Cited in ibid., 73.

abandoned the conceptual power of *Darstellung,* despite the period of his emphasis on *setzen*. For instance, Fichte's description in the *Wissenschaftslehre* of 1801 of self-consciousness in which "an eye is inserted" is but one instance of the central ocular metaphor that would dominate his later philosophy, a metaphor which sustains the rhetorical definition of *Darstellung* as a "placing before the eyes."[35]

This rather long segue into the terms (and terminological relation) of mimesis and *Darstellung* was begun in order to show how *Darstellung,* for Fichte and much of eighteenth- and nineteenth-century German idealism and romanticism, was intended as a kind of cultural translation of the Greek problem of mimesis. For our study of Fichte, such an analysis allows one to recognize that the crisis of mimesis, which erupted in the last period of Fichte's Wissenschaftslehre in his investigation of magnetic rapport (and his search for a "Physicirung des Idealismus") in which he hopes to contain mimesis's threat to philosophical truth, is a practice of containment and use of the power of mimesis that is not new, but in fact grounds Fichte's entire philosophical project as our analysis of *Darstellung* and mimesis suggests.

If it is not already clear, the attempt by Fichte and the romantic tradition to distinguish the *reproductive* from the *productive* aspects of representation through the distinction between *Vorstellung* and *Darstellung* is a distinction that attempts to free mimesis from imitation. Such a distinction is crucial in the analysis of Fichte's reaction to the imagination's encounter with technology. In other words, if one is to determine in what sense the productive and reproductive powers of the imagination are the source of truth (if not in a strictly philosophical, then at least in an aesthetic sense) and to what extent such powers are threatened by technology's usurpation of the power of imaging, then clarifying the sense of representation would be crucial to such an enterprise.[36] The conflict between the imagination and technology begins not merely with Fichte's concern with rapport in the last period of the Wissenschaftslehre, but with the very distinction between the spirit and the letter. It begins with the problem of writing and speech, through the discourse of the spirit and the letter, which we take up explicitly in the next chapter. Nevertheless, for the moment we must continue our attempt to understand how mimesis could free itself from the stigma of imitation and function as an aesthetic power of truth. This discussion of mimesis in terms of *Darstellung* is

---

35. Ibid., 74.
36. I discuss the Fichtean imagination in greater detail in Chapter 5.

essential to understanding Fichte's later metaphysics of imaging. I will argue in detail in the later chapters that Fichte's later metaphysics of imaging conceives the imagination precisely as a revelatory structure through which the individual, as a transparent structure, acquires access to the truth of absolute Being. It is for this reason that it is crucial that mimesis be distinguished from imitation. To anticipate what is to come, without being caught in a web of details, it is sufficient to state then that Fichte's entire attempt to distinguish the appearance of truth as mimesis from the imitative structure of technology will turn on his success in distinguishing degraded imitation from the imagination's revelatory power as a productive, mimetic truth. Fichte's desire to distinguish the technological power of imaging from its true source in the imagination will depend upon his ability to differentiate between appearance in terms of mimesis and imitation. If such a distinction should prove untenable, then truth itself could fall prey to the reproductive power of technology.

Affect

The value of such a discussion, in recognizing that truth does not stand as an absolute immutable form but rather, anterior to such a structure, arises from a productive revelatory mimesis, is extremely important for the issue of self-consciousness. Although, largely following Havelock, it was concluded that Platonic truth is aligned with self-conscious subjectivity as a consequence of the development of the technology of writing, writing's mimetic function points to philosophy's own limit condition to suggest that the revelatory structure of the phenomenologically conceived lived truth delimits subjectivity as a self-affective self, much like Fichte's own demarcation of an original self. In other words, in the same way that the epistemological truth of philosophy demanded self-conscious subjectivity, so too does revelatory truth, the aesthetic truth of mimesis suggests a subject anterior to the epistemological subject as a self-feeling self.

The dual nature of mimesis as both revelatory truth and Plato's account of it in terms of imitative rhetoric returns in a most unique way with Fichte's rendering of his metaphysics of appearance precisely in terms of affect. In *Anweisung des seligen Leben*, he queries, "What is the body? I say: body is the affect of Being" (*FW* 5:498). Or, further, he describes affect, or feeling itself, in ontological terms. He writes that feeling is the "feeling of Being as Being" (*FW* 5:498). This discussion of mimesis in terms of both truth and rhetoric, revelatory truth and affect, establishes a foundation essential for understanding the dual nature of Fichte's conception of our phenomenological and affective relation to absolute Being.

3

Spirit and the Technology of the Letter

Between the Spirit and the Letter

Late eighteenth-century German thought revived and transformed the classical debate surrounding mimesis and imitation. It did so primarily through the discourse of *Darstellung* and *Vorstellung*. Now *Vorstellung* designated a representation, a product of the reproductive imagination, that like *Nachahmung*, was a type of copying or a degraded form of imitation. By contrast, *Darstellung*, like mimesis, arose as a productive creation, a creation whose liminal nature appeared as a representation that nevertheless disavowed its own representational status in order to present or at least indicate a truth that was somehow beyond it.[1]

---

1. This, as we will see, is the intent behind Fichte's own conception of the transparent I (*das durchsichtige Ich*) in the last period of the Wissenschaftslehre. Here the finite self is in a state of

It is not truth in some abstract sense that is at issue. What is at stake in this long-running debate on representation is the very mode by which we apprehend ourselves and our world. After all, the structure of representation describes the very mode of the constitution of consciousness. And representation and consciousness arise out of our unique interaction with and comportment to the world. From these determinate material practices arises a sphere of interaction, a lived world. And it is this activity, the activity of the Fichtean self-active ego, which generates, for consciousness, space and time. Here, Fichte's descriptive account of the ego in terms of activity reveals an approach that is phenomenological in nature.[2] What I mean by this is that Fichte's ego is not defined as a concept, but rather emerges in its development and activity. In fact, an entire worldview stems from such an understanding. Our own interaction within the social horizon of our world exhibits a founded/founding relation to that world such that our practices both give shape to and in turn are shaped by that world. For instance, although I have been arguing that technology is a product of human literary consciousness, or in short, written language, that product itself, technology, gives structure and shape to consciousness through its technique for the manipulation of time and space. What we need to understand in this chapter then is how our own material practices, techniques, and technologies, which generate space and time, ultimately articulate consciousness, self-consciousness, and representation.

The power of space and time in the constitution of consciousness is in large part played out in the discourse of the spirit and the letter. One recalls, as Havelock and Ong argue, that it is the linearity of the typographic text, the letter, which is essential to the development and constitution of the autonomous self-conscious psyche. The technique of the letter establishes a

---

disavowal, and as such stands as a pure medium of the absolute. It is the mode of the coming to appearance of presentation of the absolute.

2. I understand this approach to be phenomenological in the general sense that the ego is not conceived as a substance or concept, but rather is determined descriptively as a "coming to be" in its activity. Here the ego is defined exclusively in terms of activity. Hegel also understands phenomenology in a similar manner. In the introduction to the *Phenomenology of Spirit* he describes his methodology of phenomenology as the "coming to be of science as such." Hegel, *Phenomenology of Spirit*, 15. Of course, such notions of phenomenology are not divorced from the greater phenomenological movement begun by Husserl in at the turn of the twentieth century. Nevertheless, as I noted, my references to phenomenology in Fichte do not reference phenomenology in this particular sense. I am also not the first to refer to Fichte's work in phenomenological terms. See, for instance, Janke's *Fichte: Sein und Reflexion*. Or for a more recent assessment of the role of phenomenology in Fichte's methodology, see my article "Reduction or Revelation? Fichte and the Question of Phenomenology" and those of others in the collection *Fichte and Phenomenology*.

temporal iteration, which is the condition of possibility of self-consciousness. Yet if time is a function of the letter, what then of space? The corollary proposition is that spirit is somehow a function of space—but how?

Spirit arose in the breath of life, in the poetic incantations of oral traditions in which the time-ordered consciousness of finite subjectivity gives way to that pre-individual moment of pure activity, to that pure movement of spatialization and space. We are presented then with the analogy: spirit:letter :: space:time. Whether such an analogy holds will be determined in the course of this chapter. The stakes of such an analogy, however, are as follows.

In this chapter I argue that our larger concern with the materialization of spirit in the age of technology is first visible in Fichte's concern with spirit's relation to the letter. This debate between the spirit and the letter stands as an important precursor to the crisis of spirit in the age of materialism because it articulates the macrocosmic social consequences of the decline of spirit (i.e., the crisis of intersubjectivity) at the microcosmic level of the constitution, development, and transformation of consciousness. Here Fichte's development of the discourse of the spirit and the letter can be understood as a continued meditation on mimesis and imitation, or in his language, as a problem of *Darstellung*, that stands fundamentally as a critique of representational consciousness. And this critical analysis of representational mediation cannot be easily separated from the very problem of technology—for instance, the technology of the letter.

Although I have explained the theoretical problem of representation, I have not yet made explicit the relation between (on the one hand) prereflective and reflective self-consciousness, and (on the other) the discourse of the spirit and the letter. In this chapter this connection will be made explicit through a discussion of the constitution and representation of the self in terms of space and time. In this chapter, I will argue that the communicative power of spirit arises out of the spatializing properties of the original ego (conceived in terms of *Tathandlung* and *Darstellung*),[3] whereas, by contrast, the letter and its linear, sequentially ordered structure mark the constitution of time and representational (*Vorstellung*) consciousness. In this sense, Fichte's definition of the ego as act defines a spatializing activity that deter-

---

3. Spirit itself, as a "primal sublime," stands beyond both space and time, and representation as such, whether conceived as *Darstellung* or *Vorstellung*. Nevertheless, like the problem of mimesis and imitation, I will be arguing that to the extent that *Darstellung* and the spatializing properties of the self-active ego precede representational time consciousness, these presentational forms remain truer communicative forms of spirit.

mines the ego as an original communicative power of spirit anterior to the self-reflective, representational ego of time consciousness.

The analogy spirit:letter :: space:time was not totally foreign to idealism, or eighteenth-century German thought more generally. For instance, in 1766 Lessing argued in his work *Laocoön* that literature stood as the preeminent art of time, and painting the preeminent art of space.[4] As with the debate surrounding the spirit and the letter, human communication was to be understood on the model of aesthetic production. As a result the exposition of space and time was conceived by Lessing almost exclusively in terms of the question of aesthetic form. Thus, while Lessing explained that the generative power of artistic communicative forms gives shape to consciousness—a consciousness whose order, and thus mechanism, can be described in terms of an aesthetic of time and space—Lessing did not yet explicitly thematize the role of time and space in the constitution of consciousness. It was not until 1781, with the reversal of the "Copernican revolution" brought on by the Immanuel Kant's *Critique of Pure Reason,* that space and time would be accorded an explicit role in the determination of cognition. Space and time for Kant were not aesthetic in Lessing's sense; rather, they stood as pure a priori forms of intuition that historically determined the very form of consciousness.[5] Yet if Kant showed the significance of space and time for the constitution of consciousness, these transcendental conditions, nevertheless, remained a priori, and therefore could not—like Lessing's description—account for the power of sociohistorically determined structures to influence consciousness. If, however, one could embrace, and thus mediate, the contributions of both Lessing and Kant in a single grasp, one might be able to understand how the artistic communicative forms of literature and art, text and image, could establish an aesthetic of time and space that orders consciousness by means of a culturally embedded transcendental.[6] As we will see, Fichte's enthusiastic embrace of both Lessing and Kant set him well on his way to such a position.[7]

---

4. Lessing, *Laocoön.*

5. That is at least in the epistemology of the *Critique of Pure Reason.* Although, as I argue in Chapter 6, the aesthetic of the Third Critique is aesthetic in Lessing's spatial sense.

6. Kant derived his transcendental categories from Aristotelian logic. When these categories were set into motion, for instance, in the work of Heidegger and Hegel, the categories were interpreted no longer as strictly logical, but rather arose as a function of cultural development, history, and time. This is what I am designating with the phrase "a culturally embedded transcendental."

7. While Fichte overtly embraced Kant, his intellectual relation to Lessing was less clear. While his own involvement in the pantheism controversy would have certainly guaranteed his familiarity

In the first period of the Wissenschaftslehre two seemingly disparate aspects of Fichte's thought seem to converge as if in part to resolve the antithesis between Lessing and Kant. First, Fichte, like Lessing, exhibits a profound concern for the aesthetic *medium* of communication. Fichte's disparagement of the written text and exaltation of rhetoric's oral traditions finds its source, one realizes, in his important distinction between the spirit and the letter.[8] In "The Difference Between the Spirit and the Letter of Philosophy" and "The Spirit and the Letter Within Philosophy," the written text is to be aligned with the imitative function of representational time consciousness, while spirit operates in a purer communicational form in the *Darstellung* of space. Here Lessing's description of the representational mode or artistic medium of communication in terms of time and space opens upon Fichte's concern with the relation between the written text and rhetoric as a question of the spirit and the letter.

Second, Fichte follows Kant—albeit critically. Here Fichte recognizes that his interest in the spirit and the letter, rhetoric and the text, as a matter of time and space, is a concern that cannot be conceived simply as aesthetic. Rather, it must be viewed as an issue of aesthetic *form* that bears directly on the structure of consciousness. Fichte's attempt to preserve and foster spirit, and his concomitant critique of the written text and its form of linear temporality, is an implicit critique of representational time consciousness. Fichte will subvert time consciousness and its literary paradigm of the text through space, a space that liberates spirit from the letter. Fichte buttresses his critique of the text and representational time consciousness by explaining what (for him) amounts to the first dynamic movements of spirit. Here, Fichte articulates a prereflective active body which achieves self-determination and self-awareness through the determination of space. Unlike Kant, Fichte does *not* consider space an *a priori* intuition, but rather a function of the activity of the self-positing I. If Kant privileges time over space, Fichte, by contrast, shows that space is primary. He shows that space arises from movement: it arises from the movement of the original activity of the self-positing I that is not readily distinguishable from space itself. The ego is identified with the activity of spatialization.

---

with the debate surrounding Lessing's Spinozism, it is at least apparent that Fichte was sympathetic to the claim Lessing made surrounding the power of time and space in the constitution of consciousness.

8. See, for instance, Fichte's insistence that the Wissenschaftslehre, like all philosophy, "can only be rescued if . . . it cannot be learned by heart" (*EPW* 212). It can only be rescued if it partakes in spirit through the living breath of rhetoric and consequently does not fall prey to the representative nature of any number of memory systems, like print.

Through the particular strengths of both Lessing and Kant, then, Fichte first came to understand the power of space and time in the constitution of consciousness. And as a consequence of this insight, and as a support to his larger project of reviving spirit, Fichte then began to develop a prereflective spatiality, a spatiality of the ego that critically undermined the time-ordered consciousness defined by Kantian apperception. In this register of representation he articulated a communicative mode of spirit anterior to the imitative connotations of *Vorstellung* and the various technologies that would constitute time consciousness, like writing, through this exploration of the limit concept of *Darstellung,* which would circumscribe a self in the spatiality of the act and convey spirit at the very limit of representation as "life."

Fichte's critique of time consciousness and the concomitant development of a prereflective ego, which defined its immediate self-awareness through space, offers us an instance or form of self-immediacy, a communicative immediacy much like the work of spirit. To reiterate, the primitive self-awareness of the original ego was not a reflective self-consciousness, but an immediate self-awareness achieved through movement and movement's constitution of space. Our discussion of space then is a discussion of this original self-awareness as a movement that arises in self-affection. Our discussion of the emergence of the original ego as self-affection through movement in space, however, cannot be divorced from the problem of communication we posed at the outset in terms of spirit. In fact, it is this communicative immediacy in self-affection that stands as an original instance of spirit.[9] Self-affection, as pure immediacy, exhibits the transparent communication that defines spirit.[10] Thus, beyond the time-ordered consciousness of discursive thinking, affection holds out the possibility (like pure spirit) of a communication outside the temporality of the letter. In what follows I come to elaborate on spirit as a problem of communication, first in the debate of the register of the communicative form of the spirit and the letter, and second, in terms of the generation of the ego, as an original communicative self-relation—made clear through a delineation of time and space.

These two aspects of Fichte's thought will be treated in the following two

9. This discourse of "affection" will be central to the articulation of magnetic rapport, both for Fichte and for this project's larger thesis.

10. The presentation of this immediacy or affection as *Darstellung,* however, reveals another limit game. As a primal sublime the infinity of spirit cannot be presented in and through a finite being without loss. Space then, produced in the spatializing properties of movement through the activity of the self-active I, is, therefore, not spirit itself, but perhaps the best possible liminal presentation (*Darstellung*) of it.

sections of this chapter. What we have yet to explore is the manner in which Fichte's emphasis on spirit (at the expense of the letter) and space (at the expense of time) finds its source in Fichte's understanding of the imagination. In the first instance, Fichte is quite clear that spirit finds its efficacy and power in the imagination (*EPW* 204). The second instance is perhaps less clear. While the imagination, as the power to produce images—images that are by nature extended—is clearly a power of spatialization, the imagination also plays an important role in time consciousness. The question then is what effect Fichte's critique of representational consciousness (through an alignment of the imagination with the spatializing act of the prereflective body) has on Kant's description of the imagination in terms of its time-generating, schematizing function. A first step toward an answer to such a question is begun in the last section of this chapter. Here our critical analysis of the role of space in transcendental philosophy and Fichte's articulation of space in terms of the act, as an original self-awareness, will begin to shed light on this aforementioned question. Through these detailed investigations of the role of the imagination in both the spirit and the letter and space and time, it should not be forgotten that our fundamental concern remains centered upon spirit, communication, and in short, the communicative space of imagination in the age of technology.

The Spirit and the Letter of the Imagination

Fichte recognized that the crisis of spirit for his age was defined by the technology of the letter and its power of imitation. Anticipating Heidegger's prophetic words about the age no longer being strong enough for spirit, Fichte writes that "the distinction of the spirit and the letter within philosophy . . . seems to be all the more necessary on account of the decisive tendency away from spirit which is characteristic of our age" (*EPW* 193).[11] Thus, it is this continuing "tendency away from spirit," facilitated by the technology of memory systems, like writing, which makes necessary the detailed distinctions between the spirit and the letter which Fichte undertakes

11. Heidegger writes: "It was not the age of German Idealism that collapsed; rather, the age was no longer strong enough to stand up to the greatness, breadth, and originality of that spiritual world." Heidegger, *Introduction to Metaphysics*, 45.

in two essays from 1794 and 1795.[12] Fichte too will attempt both to vitiate the compulsions of technology and to preserve spirit through a "talking cure," through the revival of an oral pedagogy that would come to define the very presentation (*Darstellung*) and transmission of the Wissenschaftslehre. After all, "the Wissenschaftslehre . . . cannot be communicated in any way by the mere letter, but must be imparted through the spirit" (*GA* 1:415). But how?

The problem with the transmission of the Wissenschaftslehre, like any transmission, is that "spirits are unable to affect one another immediately" (*EPW* 196). The possibility of a pure transmission, of a purely spiritual communication, thus seems barred at the outset. The problem of communication (and thus representation) becomes one of transmitting across the sensible/ intelligible divide. Thus, like the Platonic problem of the reification of the ideal, the crisis of communication for Fichte erupts as a problem of imitation writ large, because spirit, at least in part, is dependent upon its contingent embodiment, whether in the letter, or in the techniques of a given pedagogy.[13] Yet the manner in which spirit remains dependent upon the letter has yet to be explained. This section addresses the problem of the technology of the letter in light of the ideal of a purely spiritual communication. The ground for Fichte's concrete aversion to the written form in favor of the spoken breath of spirit seems to bear itself out in his account of the representational power of the imagination. As I noted, Fichte outlines the imagination in terms of its power for internal and external representation, described as productive and reproductive imagination respectively.[14] To the extent that Fichte aligns the productive aspect with spirit and the reproductive aspect with the communicative letter, he seems to set the imagination at odds with itself. Its double function or aspect, as both creative (productive) and imitative (reproductive), allows Fichte to explain the communicative exchange, all while preserving the purity of spirit. And although his distinctions may preserve spirit, we are nevertheless left with rather perplexing consequences: Fichte seems to offer us on the one hand a creative but noncommunicative aspect that, despite itself, stands as a communicative ideal (a purely spiritual

---

12. These are "Concerning the Difference Between the Spirit and the Letter Within Philosophy" (1794) and "On the Spirit and the Letter of Philosophy" (1795). See *Fichte: Early Philosophical Writings*.

13. Reification of the ideal references the discussion of the problem of self-predication in Chapter 2.

14. As we will see, the productive aspect operates by means of *Darstellung*, and the reproductive, by means of *Vorstellung*.

exchange), and on the other, a communicative and imitative aspect whose mechanical exchanges do not even truly qualify as communication.[15] These rather bizarre conclusions I believe point to a conflict between the spirit and the letter within Fichte's own explanatory enterprise. The spirit of Fichte's aspiration for a pure spiritual exchange comes into conflict, even at this level of explanation, with his delineation of the letter or actual mechanism of representation and communication. As a consequence, I believe this paradox suggests that the imagination's creative capacity cannot be so clearly partitioned from the communicative drive that is said to characterize the reproductive aspect.[16] As we will ultimately see, this dilemma of spirit involves not only the imagination; it is founded in the representational drive, a drive whose dynamic of self-activity and limitation is the very source of the Fichtean "I."

The overt dependency of spirit upon the written text intersects with its theoretical or psychological explanation of the dependency of the productive imagination upon its reproductive counterpart in the communicative paradigm of the artistic genius. In the "Difference Between the Spirit and the Letter of Philosophy," Fichte contrasts the artist, as one with the capacity to imbue the letter with spirit, with the "mere technician," who even with the greatest skill can never "produce anything but a mechanical work" (SLP 90). And in the "Spirit and the Letter Within Philosophy" Fichte attests to the vacuity of mere imitation through his description of spiritless reading and writing. For him, the process of copying, or imitation, is the very definition of an act without spirit (*EPW* 198–99). He explains, just as "the person without spirit obtains his rules from without," the individual who practices script, and "glances back at the model he is copying" "with every stroke of his pen," is devoid of spirit because "handwriting is not yet something natural for him" (*EPW* 198). While one might read this passage, emphasizing the "yet," to argue that handwriting, once natural, might eventually embody spirit, there is the suggestion, that will be still clearer in the example of reading which follows, that the capacity or choice for writing to be performed with or without spirit in fact relegates it to the status of a tool that is never fully naturalized. Or perhaps even that the very distinction upon which spirit

15. This paradox of imagination seems to reiterate the very paradox of reflection I highlighted (in Chapter 1) through reference to the work of Dieter Henrich.

16. Such a conflation is clear in the controversy surrounding the very nature of mimesis and the historically fluid nature of the terms *Darstellung* and *Vorstellung*.

differentiates itself from the letter, by means of delimiting the natural from the unnatural, is itself fraught with difficulty.

Fichte now turns to consider the act of reading and the status of the book. He points out that the often experienced tediousness of reading is the result of the difficulty of imbuing the written work with spirit (*Geist*). If Fichte then is describing *Geist* as "the vitalizing force in a work of art, and its absence *Geistlosigkeit*," then the question is how one work can convey spirit, while another cannot. Not surprisingly, Fichte tells us spirit is best conveyed through the dead matter of print in those works that are still able to manifest characteristics of the spoken word.[17] The secret is that the work must offer "not just the gift but the hand with which to grasp it." The technological medium must also embody a proper theory of reception. Consequently, in doing so, "such a work creates the spectacle and the audience at one and the same time and like the life force of the universe, imparts first movement and structure to dead matter and then in the same breadth spiritual life to that structure" (SLP 77). If it is to manifest spirit, written work must imitate speech because spirit resides in speech. The ability to convey spirit through a given conduit, like the letter, is the realm of the artist.[18] The artistic genius stands as a paradigm for Fichte precisely because his or her ability to communicate spirit by means of the letter is an ideal that would define any true communication.

True communication is possible for the artist because for the artist the body of the letter is the vessel of the spirit, whereas for many the contingent form of the letter is but the effect of an empty process of mechanization. For the artist, the contingent form of the letter is the sole means to communicate spirit. Yet while the letter is the key to communication for the artist, Fichte's description is often too eager to realize the communicative ideal: he imagines the letter stands as a transparent form or conduit for a spirit to spirit exchange. This is readily apparent in certain correspondences he has with Schiller, who takes him to task for insisting that one see the spirit of his work beyond the medium of its rather obscure, and for Schiller, often impenetrable style. When Fichte, in effect, urges that Schiller intuit what he writes, Schiller quickly reminds him of the importance of the form or letter for conveying the spirit. While Fichte hopes that his style can be overlooked in the face of

17. And it is this positive embodiment of spirit which Fichte is trying to articulate through the term *Darstellung*.

18. It is the productive power of mimesis. It is a *poesis*.

the power of the spirit of his work, Schiller remains quite critical of his style. Schiller writes: "You thrust the reader directly from the most abstruse abstractions right into harangues: this is the source of the unseemliness of your writings" (*EPW* 397). Fichte's defense again attempts to move beyond the letter and embrace spirit in its immediacy. In response, he essentially suggests that the meaning of his writings would be clearer if one could step back from the letter and return to an oral form, which he infers would allow one to more readily embrace spirit. He writes: "The main reason my sentences appear stiff is because readers are unable to read my sentences aloud" (*EPW* 395). One must reactivate the dead text through the spirit of the spoken word. He concedes to Schiller, however, that today "our public simply cannot read aloud, and one does better to set one's standards accordingly" (*EPW* 395).

Yet if Fichte sometimes ignores the importance of the medium in practice, it is nevertheless of great theoretical concern to him. His discussion of the artist's work in terms of the incarnation of spirit is an important anticipation of his later discussion of the phenomenon of magnetic rapport in terms of a "Physicirung des Idealismus."[19] While I will eventually detail the notion of a *Physicirung* in the latter portion of this book, for the moment it is sufficient simply to realize that Fichte intends the letter, if it is to truly communicate at all, to be a *Physicirung*, a physicalization of the spirit. It is the artist and his work which first exhibits the capacity for enacting this *Physicirung*, for giving spirit a material incarnation. The issue, however, is still one of *Darstellung*.[20] One might then also say that the measure of the artist is his or her capacity for the *Darstellung* of spirit. Yet communication, even in this specific sense, cannot be limited to the artist alone.

If communication is what defines community and, in effect, spirit, it would be counterintuitive to suggest that it is limited to the artistic genius. In fact, Fichte points out that the artist stands as a model for us all. The artist stands as the contingent embodiment of the universal spirit and embodies the very possibility of communication. This ideal of communication seems

19. See TaM 70.
20. The representation, materialization (*Versinnlichen*), and communication of the absolute for the artist, particularly in his or her role in the sociopolitical community, is conceived in the first period of the Wissenschaftslehre in terms of *Darstellung*. It is in the last period of the Wissenschaftslehre, in Fichte's "Tagebuch über den animalischen Magentismus," that this same problematic is expressed in terms of the desire for a *Physicirung*. My use of this latter term to describe some of Fichte's early work is done to emphasize a continuity in Fichte's concern despite some shift in vocabulary.

to turn upon a *Gemeinsinn,* a common capacity in which the artist's capacity for communication "lies in every breast, and his capacity is the common capacity [*Gemeinsinn*] of the whole species" (SLP 88). In other words, "what is posited through the essence of reason is the same in all rational individuals" (SLP 88). It is not clear at this point, however, that the *Gemeinsinn* that would be the hallmark of a rationally based theory of communication is what Fichte's account has in mind.[21] We will return to this issue as I further develop the problem of imitation in the account at hand. For the moment, however, we need to return to the question of the imagination.

Fichte offers a theoretical account of the relation of the spirit to the letter by exploring the representational power of the imagination. He begins by explaining that spirit depends upon the imagination for its task of bringing feelings to representation (*EPW* 199). In brief, the faculty of imagination carries out the work of spirit through a process of schematizing or imaging. Imagination's process of bringing feelings to representation can be further understood in terms of two distinct moments or aspects. These are referred to as the *productive* and the *reproductive* imagination respectively. The productive imagination is wholly creative: "It creates something from nothing" (*EPW* 193). The reproductive imagination, by contrast, is precisely imitative: it may assemble new combinations, but it "repeats something which was already present within empirical consciousness" (*EPW* 193). Yet what is the role of each aspect in representation?[22]

The dynamic of the imagination is perhaps most clearly understood in terms of the representational drive. Yet if Fichte's understanding of the dynamic of representation finds its origin in the drive, before we begin to develop it we would do well to understand what sort of knowledge that is, and how he arrived at it. If spirits cannot communicate with one another directly, how does one come to read the inner workings of spirit (productive imagination) or the soul? In other words, how does Fichte come to postulate the notion of the drive?

Fichte makes clear in "On the Spirit and the Letter Within Philosophy" (1795) that our very understanding of the imagination is itself an act of

21. I am, of course, alluding to the work of Jürgen Habermas. Nevertheless, it is the intent of this entire project to show that what might first appear as a rationally based theory of communication is in fact subtended by an affective social bond. As we will see, this is undertaken primarily through a rereading of intersubjective recognition on the order of an aesthetic of affect. See especially Chapters 5 and 6.

22. Fichte generally associates the productive imagination with *Darstellung,* while the reproductive imagination engages in a copying that is typically described as *Nachahmung.*

interpretation and representation by which we must read the spirit, soul, or in this case the representational drive, by the effect it produces. It is through the effect that we are to infer the topography of the imagination. Fichte explains: the "drive manifests itself in the effect it produces. From this we return to the cause in the self-active subject, and only in this way do we arrive at the idea of the nature of this drive and a recognition [*Erkenntnis*] of its laws" (SLP 80). If it is from the effect of the drive that we infer the cause, as the self-active subject—insofar as the drive is the ground of all self-activity, and the subject itself is defined in terms of act—then the very idea of the subject in Fichte appears unconscious (*unbewußt*).[23] While we will take up the issue of the unconscious subject again in the next chapter, it is nevertheless important to recognize what is at issue here for our discussion of the imagination. What we witness in the discussion of the drive is a conflict between the actual path of Fichte's epistemic methodology and the epistemic ideal that is placed before us. Through his demarcation of the imagination according to productive and reproductive aspects, Fichte hopes to delineate a spirit (to the extent it is aligned with the productive imagination) free from the contagion of the reproductive imagination's imitative function. Once again mimesis is to be freed from imitation—and spirit freed from the letter—by distinguishing *Darstellung* from *Nachahmung, Nachbildung,* and *Vorstellung.*

Two problems remain. The first is that since, as Fichte conceded, immediate spiritual communication is impossible, his very knowledge of the topography of the faculty of imagination is the result of his inferences from the effects of the drive. He himself, however, has no immediate knowledge of the imagination. Like the hermeneutic of symptoms by which Freud delimits the topography of the unconscious, Fichte too can only arrive at a figuration of the imagination as the result of an inference from the effects of the drive.[24] As such, the problem of imitation enters the very heart of his account of the imagination and its representational drive. Second, as I will detail in the following chapter, to the extent that spirit is fundamentally involved with the idea of communication, it is not clear whether Fichte is justified in aligning spirit only with the productive imagination and not also with the reproductive aspect—the key element in external communication. Or in the discourse

23. *Unbewußt,* of course, is literally "non-conscious," "not-conscious," or perhaps even "preconscious." Strictly speaking, it does not designate the Freudian notion of the unconscious.

24. In this sense Freud's hermeneutic of the unconscious can be understood to have much in common with transcendental philosophy.

of mimesis, it is not clear whether *Darstellung* can be freed from *Vorstellung* (or *Nachahmung*), or mimesis from imitation.

Fichte aligns spirit with the productive imagination and distances it from the imitative and communicative representational role of the reproductive aspect in order to give a metaphysical ground to his concrete distinction between the spirit and the letter—a distinction intended to save "spirit," admittedly an already mediated spirit—from the rise of mechanical reproduction. Fichte's notion of the drive would seem to undercut his attempt to separate spirit from its essentially communicative vocation. First, according to Fichte, the drive is not truly divisible; rather, it is an *in*divisible originary force (*Grundkraft*) (SLP 80). Second, since our very conceptions of spirit, soul, or the "I" are but effects whose source we designate as the drive, the very idea of these notions all amount to a self-active force or activity whose action is essentially communicative (*EPW* 200).

Once again then I begin to address the relation between the productive and reproductive imaginations by articulating the representational drive. Although Fichte outlines three drives—(a) the knowledge drive, (b) the practical drive, and (c) the aesthetic drive—I reiterate that they are at root "one indivisible originary force [*Grundkraft*]" (SLP 80). The aesthetic and the practical drives stand as correlates to the productive and reproductive aspects of the imagination respectively. The aesthetic drive is concerned merely with representation (*Darstellung*) as such, "exclusively for the sake of its determination" (SLP 81), while the practical drive (via *Nachbildung*) "seeks to bring forward something that corresponds to it in the sensible world" (SLP 81). As a consequence Fichte will distinguish the respective roles of the productive and reproductive imaginations by distinguishing the "aesthetic drive" from "aesthetic images."[25] He explains that, with the aesthetic drive, the original moment of representation "terminates completely in the mere projection [*Entwerfung*] of images within the soul," while the practical drive focuses on the presentation of aesthetic images and "sets up imitations [*Nachbildung*] in actuality" (*FW* 8:281).[26] In other words, the productive moment of imagination brings mute feeling to consciousness through the process of imaging. Its sole task is representation for the sake of representation. These translations of feeling, as images projected upon the soul, however, cannot yet be communicated. Communication, the process by which this originary representation

25. John Sallis emphasizes this point in *Spacings*.
26. Cited in Sallis, *Spacings*, 31.

is externalized, is the task of the reproductive imagination. *Darstellung* must be converted to *Nachbildung*. Yet is this distinction justified? Can representation justifiably be demarcated into distinct internal and external dynamics? Or is the drive to representation continuous such that one is compelled to externalize that which was internally represented? After all, isn't this the fundamental difficulty with mimesis?

Despite Fichte's assertion that spirits cannot communicate directly, he was hard put to let go of the ideal of a pure spiritual communication. Fichte, in effect, attempted to ignore the body of the letter and to enact a spirit to spirit transmission. On July 2, 1795, Fichte wrote to Reinhold: "What I am trying to communicate is something which can neither be said nor grasped conceptually; it can only be intuited. . . . I advise anyone who wishes to study my writings to let the words be words and simply try to enter into my series of intuitions" (*EPW* 398). But while Fichte hoped for a immediate spiritual communication through intuition, beyond the letter, some of his readers, like Schiller (as I noted), wished that he would pay more attention to his own artistic and stylistic form. What Fichte's remarks further suggest is that one cannot relegate the communicative role to the reproductive aspect, and the creative to the productive, but rather that a fundamental communicative drive underlies the entire faculty of the imagination. How else could one explain Fichte's intent to communicate beyond the letter? In fact, against the letter of his own writings Fichte seems to suggest that spiritual representations (*Darstellung*), even without the clothing of contingent imitative forms, are precisely "for the purpose of communication between spiritual beings" (*EPW* 199).

A further hint that spirit or the productive imagination itself strives to communicate is apparent in Fichte's profoundly curious discussion of what amounts to a spiritual copying. It is also our first hint (as Plato realized) that imitation or mimesis is perhaps somehow caught up with the very condition of possibility of communication. Fichte begins by noting that "there is in all of us the drive to make the people around us as like ourselves as possible, and to duplicate ourselves in them as closely as we can" (SLP 88). What this statement suggests is something much more dangerous than perhaps Fichte realizes. What Fichte would not readily allow was the mechanical act of copying or imitation across the sensible/intelligible divide, between the spirit and the letter. Yet one would imagine that what Fichte attempted to avert was the danger of mechanical imitation, which culminates in its most extreme form in the notion of the "Platonic double," or "pure mimos," which we spoke of earlier. Ironically, however, Fichte attempts to maintain a distinction between the spirit and the letter, all while apparently embracing what for Plato was the ultimate horror, which these more benign forms of mechanical imitation only anticipated: the political contagion of the pure *mimos*.

As we will see, Fichte, in effect, reinterprets and rereads the moment of Plato's most direct encounter with mimesis—his encounter with the pure mimos—in *The Republic,* but without recognizing it as a danger. In fact, Fichte suggests that this doubling is akin to flattery. Thus, although Plato was perhaps not so optimistic about mimesis, Fichte expresses a different outlook. Surprisingly, Fichte's caveat—that duplication serves a higher cause, and is in fact an expression of nobility—seems to suggest that he believes mimesis can in fact be free from imitation. He even further hints that it would be unjust not to engage in such mimetic communication. He writes: "This is all the more so when we are justified in this aspiration because of our own higher development [*Bildung*]. Only the unjust egoist wishes to be the sole of his kind . . . but the noble person would like everyone to be like him, and does everything in his power to bring it about" (SLP 88). His final argument here is, in effect, that one ought to accept this mimesis as the essence of communication, because it is natural; nature proceeds in the same manner. Fichte continues: "This is how it is with the darling of nature. He would like to see his own darling image reflected back to him from all other souls. Thus he imprints the mood of his spirit into a physical form" (SLP 88).

This call for a *Physicirung,* for an imparting of spirit in its physical form, continues to affirm this notion of the medium that we have been speaking of with the terms mimesis, *Darstellung,* and "subtle matter's" *Physicirung.* In other words, Fichte's embrace of a purely spiritual communication can do no better than to embrace a mimesis freed from imitation. Thus, it is not truly a pure spirit to spirit communication, but rather the attempt to articulate a pure aesthetic medium that could function beyond the baser forms of imitation. Yet to insist on the ideal of a pure spirit to spirit communication, all while recognizing its impossibility, is to wish for a pure communicative medium and, in effect, invite disaster. Fichte, once again, is caught up in the divide between sensible and material worlds. And his search for a pure medium, a material form with immaterial characteristics, leads him not only to insist on the possibility of a mimesis freed from imitation, such assumptions lead him to magnetic psychology's notion of subtle matter, and ultimately to the notion of a "Physicirung des Idealismus," through which he will attempt to defend his later metaphysics against the tide of materialism.

Still, we need to recognize that Fichte's demand for spirit to spirit communication largely materialized in any number of figures of immediacy—for instance, in the form of the productive imagination and the self-active ego. Thus, as we will see, through what could be understood as a phenomenology of immediacy, Fichte attempted to articulate a form of (re)presentation, as *Darstellung,* anterior to the constitution of time consciousness. He did so by showing that time consciousness is ultimately grounded in space, in the spatializing power of the self-active ego.

4

The Spatial Imagination

Affect, Image, and the Critique of Representational Consciousness

Communication and the Problem of Space

If communication is to take place, the intelligible realm must somehow be communicable across the divide of the sensible world. Spirit must be given a material form. While the discussion of the spirit and letter arose from a predominantly aesthetic paradigm, the issue it engages is certainly not limited to an aesthetic of communication. In fact an investigation of the sensibilization (*Versinnlichen*) of spirit and its communicative power of the imagination, which the question of the spirit and the letter had merely broached, is fundamentally an inquiry into the constitution of consciousness and the world-constituting power of the original ego. In fact, we must critically develop the extent to which the *Darstellung* of the ego and the materialization of spirit each intersect as a part of the same project of *Versinnlichen*. In other words, in what sense can Fichte's analysis of *Darstellung* and his call for a "Physicirung des Idealismus" be thought of as part and parcel of the same process of *Versinnlichen*? Yet while the self is ultimately confronted with the problem of communication in the register of its aesthetic form, the constitution of this representational consciousness arises through the very same conflicted dependence that defined communication at the interstices of the spirit and the letter. Except now, with the explanation of consciousness and the ego, the terminological register has shifted. While spirit was that original form which was dependent upon the letter for its appearance in the genetic deduction of consciousness, the space-constituting movement of the original self-active ego will depend upon time for its manifestation. In short, the original ego, which arises as a function of space, is ultimately dependent upon time consciousness for its discursive appearance.

Just as the issue of mimesis was broached in the discourse of the spirit and the letter, *Darstellung* and *Vorstellung,* so here too is the very same problem of representation articulated as an issue of the constitution of consciousness through a discussion of the relation of space and time. From the same impetus that led him to read the spirit of Kant against its letter, Fichte continued to undercut Kant's singular account of the logical ego of representational time consciousness by developing its conditions of possibility through his articulation of the constituting power of space. Just as the letter is reinterpreted through its spirit, so too will the logical ego, identified with time consciousness, be reformulated by developing its origin in space.

Like spirit, space is original. But also like spirit's dependency upon the letter, so too is space dependent upon time for its discursive representation within consciousness. Thus, the problem of the antinomy between sensible and intelligible worlds remains. In the rubric of space, we might say that mental space must be articulable across physical space—but how? Since Descartes, the transcendental subject had been delimited as a mental entity, *res cogitans,* by analogy with and opposition to the physical space of the *res extensa.*[1] If the subject defines itself by analogy with and in opposition to the *res extensa,* however, the transcendental subject and the very project of transcendental philosophy quickly concludes that the physical space of the object is in fact a function of mental space.[2] The problem with this final step,

1. See Lefebvre, *Production of Space.*

2. This is the fundamental thesis of the transcendental turn. Space for Kant "belongs only to the form of the imagination, and therefore to the subjective constitution of our mind" (*CPR* B38). Mental space would subsume physical space, and consequently be conceived as absolute (see Lefebvre, *Production of Space,* 1–3).

which reached its apex with Kant, was that—as Fichte was well aware—the absolutizing of space would lead to an insurmountable abyss between the mental space of epistemology and the physical space that is part and parcel of our social world.[3] This crisis of space, opened up in the abyss between the intellectual and the sensible, is part of the larger crisis of communication—a crisis of spirit—which we have been trying articulate.

Communication designates a social space that would bridge the opposition between sensible and intelligible worlds. Nevertheless, it is perhaps precisely this opposition or distance which makes communication as such possible. After all, the very notion of communication assumes space, like the distance between sender and receiver. If either identity or distance were absolute, communication would either be unnecessary (in the first instance) or impossible (in the second). This paradigm of communication then appears to mirror the part-whole relation of the Absolute to its particular manifestations.

If communication is to function as a bridge between mental and physical space, and thus make possible the communicative space that would define the social, one needs to recognize that its synthesizing power lies in its source—in the imagination. The imagination constitutes communication, thereby mediating the sensible and the intelligible through its schematizing function, through its power to produce images. Yet what sort of space does the imagination itself construct? What is the productive power of imagination that makes the space of communication possible? After all, isn't the image itself that which is visually extended?

The fact that the image is a percept of pure extension offers us a clue. As a power to produce images, the imagination exhibits the power to produce a certain sort of space. It produces a communicative space that would constitute community through the capacity of the image to bridge the sensible/intelligible divide. An account of the role of the imagination in communication then must begin with its power to give figuration to space through the image. It does so through schematization. This synthesizing function of the imagination, however, is a structure of representation that constitutes not only objects but self-consciousness itself. Yet is consciousness, like the image, a function of space, or of time? If, as we gleaned from our discussion of Lessing, the linearity of the written text is communicated to consciousness

3. Lefebvre writes: "The quasi-logical pre-supposition of an identity between mental space (the space of the philosophers and epistemologists) and real space creates an abyss between the mental sphere on the one side and the physical and social sphere on the other" (*Production of Space*, 6).

across time, and the image is defined by an instantaneous spatiality, then one might suggest that Kant's privileging of time in the constitution of consciousness is an effect of the literary tradition (i.e., the written text).[4] Could Fichte's emphasis on prereflective activity then stand as yet another attack on the printed text and its aesthetic of time consciousness? If so, the battle between the spirit and the letter, the written and oral traditions, can also be understood to be fought primarily on the terrain of consciousness. Fichte's battle against the tide of written texts, through his insistence upon an oral tradition, would begin by an at least partial overturning of the privileging of time in the constitution of consciousness through a revalorization of space.[5]

One recalls that Fichte's awareness of G. E. Schulze's critique of Reinhold's description of a purely representational consciousness led him to articulate a pre-representational consciousness.[6] And if representational consciousness was determined by time, then Fichte's prereflective active body would achieve self-determination and awareness through space. In what follows I argue that Fichte's emphasis on the significance of the prereflective spatiality of the body tends to attenuate the Kantian privilege of time in the constitution of consciousness. We begin then with a detailed analysis of the role of space and time in Kant's First Critique.

The Critique of Representational Consciousness: The Priority of Space in Imaginative Synthesis

Our inquiry into space and time is for the sake of illuminating the dynamic capacity of the imagination as a faculty of communication. The imagination was, after all, said to be not only that faculty which could bridge the sensible/ intelligible divide, but as such, the very source of spirit. Spirit, we recognized in the previous chapter, was not readily or necessarily conveyed by the letter. In fact, Lessing's identification of the written text with an aesthetic of time, and by extension time consciousness, would lead one to tentatively conclude

---

4. In fact, the more general suggestion that Kantian categories are historical manifestations rather than cognitive absolutes has already been made by both Hegel and Heidegger. Such claims support a historicized reading of Kant that suggests that his work—like Plato's—is the effect of literary consciousness.

5. Fichte again shows his annoyance with books. See *FW* 8:97.

6. See Chapter 2, or "Review of *Aenesidemus*" (*EPW* 65).

that spirit is not of time, but perhaps, in a certain sense, a function of space.[7] Time is an expression of our finitude. It is an expression of finite consciousness. Yet if space designates that impossible sphere of a purely spiritual communication, then perhaps like the communication of the spirit through the letter, if real communication is to occur, it (space) must be communicated—at least to some extent—across time.[8] This is not to say, however, that space is merely reduced to time. One need not lose spirit in the process of mechanization. Mimesis need not be reduced to imitation.

Fichte's critique of Kant's transcendental ego—as a logical subject, generated through the unity of apperception—begins with his own account of the self-active subject. This original self-active self arises in movement, a movement whose anteriority to the Kantian ego (manifest as the effect of time consciousness) tends to undercut the Kantian representational subject. Fichte's emphasis on the imagination, as an activity and movement, stands as an important corrective to Kant's account of the imagination's process of schematization as a process of temporalization that (for Fichte) does not truly unite sensible and intelligible, internal and external worlds. The imagination is fundamentally a power of communication for Fichte because its oscillating (*schwebende*) action stands as a communicative relay between infinite and finite worlds.

For Kant, both time and space are pure intuitions and as such form an intrinsic part of subjective sensibility. Thus, while time is said to describe the form of "inner sense," and space the form of "outer appearance," both are a priori pure intuitions. According to Kant, they are "the pure forms of all sensible intuition" (*CPR* A39/B56) that define the very mode of our sensibility. Whether Kant's distinction between time and space, and inner sense and outer appearance, is justified, however, is not clear. In particular, it was Fichte's dismissal of the "thing-in-itself" that put this Kantian distinction into question.

If space, as the form of outer intuition, had served as the conduit of sense impressions from the thing-in-itself, then with the demise of the thing-in-itself outer sense was no longer simply given, but also produced (*WLnm* 240–41). Space then, like time, is something that finite consciousness gener-

---

7. This, as we will see, is only partially true. Spirit, as the primal sublime, is beyond both time and space. On the scale of immediacy, however, read in terms of our experience of spirit, space can nevertheless be understood as a more original form than time.

8. In the language of mimesis this is equivalent to claiming that mimesis to some extent must partake of imitation.

ates in its process of apprehension. Apprehension is not simply abstract cognition, but arises in activity. And space is generated in activity (just as time is generated in activity and apprehension). For Fichte, the Kantian distinction, which privileges time in the imaginative synthesis, harbors within it several presuppositions that serve to buttress the Kantian account of the transcendental unity of apperception.

Although Fichte's critique of the representational consciousness proffered by the Kantian account of the transcendental unity of apperception certainly arises through his articulation of the concepts of self-activity and immediate self-awareness, at the most fundamental level this critique proceeds through a revalorization of space as the ultimate foundation of time. For Fichte, activity, as power of spatialization, forms the foundation of time consciousness. Fichte argued that the Kantian I, as a representational *ego cogito,* literally appears on the order of phenomena, and as such arises as the effect of time. By contrast, the Fichtean self is an activity anterior to time consciousness, and is identified by Fichte as a noumenon, as a pure force, whose process of spatialization is preconscious, nonconscious, or unconscious (*unbewußt*).[9]

From Fichte's perspective, Kant's privileging of time in the schematization of the imagination and the constitution of consciousness is the result of his own particular understanding of the imagination. For the Kant of the First Critique the imagination remains a maidservant of the understanding, while for Fichte it is the imagination that is the source of spontaneity and the guide to the understanding.[10] For Kant, then, it is through the process of schematization, or imaging, that the imagination would unite the sensible and intellectual components, thereby making experience as a whole possible. For Kant this schematization constitutes time, and is thus bound to the intuition of time—understood as "inner sense." For Fichte, however, the sensible and intelligible worlds are united through the concept of action, through the intellectual intuition of a self-active self, whose sphere of determinability is given in and through space.

According to Fichte, Kant never convincingly explained the intersection of the sensible and the intelligible worlds, the relation between inner and

9. The idea of a noumenal self does exist in Kant. He writes, "The soul in transcendental apperception is *substantia noumena*" (Reflexion C001, *GS* 18:420–21; cited in Allison). And Henry Allison points out that although this is Kant's official position, it is also the source of "two distinct and incompatible doctrines about the subject of apperception and the noumenal self." We are, however, less concerned with Kant than with Fichte's own particular interpretation of Kant. See Allison, *Kant's Transcendental Idealism,* 286.

10. See Makreel, "Fichte's Dialectical Imagination."

outer sense. These determinations (according to Fichte) remain pure abstractions divorced from any real practical instantiation. Fichte began to question the Kantian articulation of time by questioning the very validity of the Kantian categories. In the *Wissenschaftslehre* of 1794 Fichte states: "I am aware that he [Kant] by no means proved the categories that he set up to be conditions of self-consciousness, but merely said they were so: that still less did he derive space and time as conditions thereof" (*FW* 1:478). To correct what he saw as Kant's failure, Fichte united the sensible and intelligible worlds through the concept of action. If, as Fichte shows us, action is what constitutes space, and this movement in space is what ultimately generates time, then the key to the unity of time and space, inner and outer sense, and the sensible and intelligible worlds, lies in Fichte's understanding of the "act."

Now if Kant, I am arguing, subsumes space to the model of time—as generated in the process of schematization—Fichte reversed this emphasis by upholding space and its sphere of action as the founding paradigm. If we look at Kant more closely, we realize the transcendental schema of the imagination unites the intellectual and sensible aspects by applying a category of the understanding to appearance through the transcendental determination of time. Since the inner sense cannot realize the entire contents of outer sense (sensible appearance) in a single instant, but rather does so in consecutive instants, piecemeal, the synthesis of sensible appearance takes place across time. In fact this synthesis, according to Kant, generates time. He writes that this synthesis or unity is "due to my generating time itself in the apprehension of the intuition" (*CPR* A143). Kant's description of the synthetic power of the imagination is a phenomenology, a phenomenology that explains the infinite appropriating acts of a finite consciousness through the determination of time. As finite beings, our apprehension of phenomena generates time. Yet, as our reading of Fichte will make clear, Kant explains this phenomenological act of the synthesis of the imagination by privileging the discourse of time over space. According to Kant's system this would make sense, since it is space, as "the a priori condition only of outer appearances," which needs to be explained in terms of inner sense, or time.[11] The schematization of space is the translation of space into time. Kant explains the thought behind his translation of space into time as follows: "The pure image of all magnitudes [*quantorum*] for outer sense is space. . . . But the pure schema of magnitude [*quantitatis*], as a concept of the understanding, is number, a

---

11. Kant writes, time "is nothing but the form of inner sense" (*CPR* A33/B50).

representation [*Vorstellung*] which comprises the successive addition of homogeneous units. Number is therefore simply the unity of the synthesis of the manifold of a homogeneous intuition in general, a unity due to my generating time itself in the apprehension of the intuition" (*CPR* A143). Space is translated into the discourse of time through number, through the successive counting of spaces, whose iteration marks time. Time's origin in space, however, was no surprise to Kant. He writes: "Thus our concept of time explains the possibility of that body of a priori knowledge which is exhibited in the general doctrine of motion, and which is by no means very unfruitful" (*CPR* B49).[12] Time is erected on the model of space, and space is defined in terms of motion. This observation by Kant was not lost on Fichte. In fact, it forms the foundation of Fichte's very critique of (Kantian) representational consciousness (insofar as it is a function of time). It is a critique that will redefine thought and consciousness in terms of activity (*Tathandlung*). Yet if, according to Kant, the manifold of intuition is given order through time, we still have neither determined nor explained the form of that self-identical consciousness to which the flux of the manifold is to be presented.

The Double Ego: Fichte's Critique of Kantian Apperception

The Kantian account of the transcendental unity of apperception presents two central questions for Fichte. The first asks how the manifold of representations is linked together, that is, unified. The second demands a self-referential something to which the fluctuating manifold can be presented. In brief, Kant fulfills these difficulties (in the first instance) by linking the manifold together through time and (in the second instance) by presenting the manifold before a logical subject (B132–34). Fichte, however, is not completely satisfied with Kant's response. He objects to an account of time that defines the logical subject as that representation which accompanies all other representations, by bringing the manifold to unity before the subject. This description is perhaps better known as the transcendental unity of apperception.

Fichte sees the "aggregate of feelings" that would make up the manifold to be linked through an interdependence that makes up a series. According to Fichte, "the feeling of the manifold in the relationship of dependence"

12. Cited in Ricoeur, *Time and Narrative*. Ricoeur, also following Fichte, sees Kant as deriving time from the model of space. Ricoeur writes: "The transcendental exposition of time is constructed exactly on the model of that of space." Ricoeur, *Time and Narrative*, 3:46.

is realized in the act of schematization—and schematization is a temporal succession. Yet while time is the form of inner sense that makes sensible intuition possible, this process of schematization occurs exclusively at the level of phenomena.

The temporal series continues through the I's repeated acts of self-determination. Yet for Fichte, the original I is not a phenomenon but a noumenon (*WLnm* 400, 408).[13] Here the I, as pure energy or act, does not itself occur within time, but through the act of determination or intentionality gives unity to the manifold. By determining what is in time—that is, through the synthesis of the manifold—the I itself becomes an objective instant in time. The original unrepresentable self—or in Fichte's terms, the I as noumenon—is given to itself through the process of schematization by which its entry into time makes it possible to appear before itself as a sensible intuition or object. What appears before the original I then is the I as phenomenon: the representational *ego cogito*. As a power of self-determination the I is beyond time, but in the act of determination the I appears as a part of the temporally conditioned synthesis that brings the manifold to sensible intuition (*WLnm* 269). Fichte writes, "I observe myself within time, but I am not in time, insofar as I intuit myself *intellectualiter* . . . as self-determining" (*WLnm* 280). Fichte continues by explaining that this I, as force or act, is only "a single act of willing; and whose original unity is what makes the series possible in time. Since the I first becomes aware of itself through the temporal series, and sensibilized, it is clear that 'time is the form of intellectual intuition'" (*WLnm* 281).[14] The repetition of the I's self-awareness in intellectual intuition is the first sensible index of permanency that both makes succession visible and offers a first account of the role of the imagination in memory. Yet if time is the process by which the activity of the will is first brought to sensible intuition or appearance, it is this original act or sphere of action that will be described in terms of space. The imagination then is movement in its truest sense: it functions as a pure movement, as the dialectical relay between the phenomenal I as the effect of time, and the original, noumenal I, which is aligned with space—the absolute space of spirit.

While Kant's description of time and space, particularly in the Deduction, is notoriously obscure, Fichte takes Kant to mean that time and space are

13. Fichte asks himself, "But is this latter way of looking at the I [as a power]. . . , {just another way of looking at the first, original noumena}?" (*WLnm* 408). See also *WLnm* 400.

14. Intellectual intuition thus appears in the form of time. It is a form of self-immediacy. But, as we will see, self-feeling is anterior to time and intellectual intuition.

inseparable modes or aspects of the I. Time designates the imaginative synthesis that constitutes the appearance of the I under the intellectual aspect, while space is a description of the I from the perspective of the body (*WLnm* 396). With the Kantian absolutizing of space, the body remains abstract and is ultimately intellectualized through the privileging of time over space. Time generated in the imaginative synthesis then is the privileged mode of the representation of the self. Kant was well aware of this paradox. He explains: "This is a suitable place of re-explaining the paradox which must have been obvious to everyone in our exposition of the form of inner sense: namely, that this sense represents to consciousness even our own selves *only as we appear to ourselves, not as we are in ourselves.* For we intuit ourselves only as we are inwardly affected" (*CPR* B153; my italics). It is clear that a self anterior to representational consciousness is also a self anterior to time consciousness. Since our represented self is itself subject to the conditions of appearance, and in fact manifests itself as appearance, we cannot appear, Kant notes, "as we are in ourselves." The "I" as noumenon—which Fichte seeks—is literally unconscious. Fichte identified this noumenal I with a force, a "sheer doing," or act, that arises by analogy from movement—from the very spatial paradigm that Kant subsumed under time (*WLnm* 405).

This introduction to the critique of representational consciousness through the notion of the original ego opens us to Fichte's idea of intellectual intuition; but it is a stage that we must work toward. We will begin with the first act of the ego and its originary self-awareness in self-affection as it emerges from space. Even the original unity, we shall see, is synthetic to the extent that its original act is itself identified with the synthetic power of the imagination.

Space, Imagination, and the Original Ego

In the first period of the Wissenschaftslehre Fichte begins to develop the notion of an original ego: the ego defined as act. The schematization of the imagination in Kant synthesized diverse moments into a single whole, thereby giving unity to consciousness. Fichte reworks this Kantian temporal synthesis, however, as a spatial synthesis undertaken by the self-active ego. For Fichte, if space is the field of movement or action, then the synthesizing function of the imagination gives unity to the ego in unified activity. The imagination makes action possible in two ways: in the first instance, as a power for synthesizing space in a field of unified action known as the ego; and in the second instance, by dialectically uniting the divergent valences of the ego—striving and limitation—which define the ego as act.

In order to make clear Fichte's understanding of the original ego (prior to the *ego cogito*), we must further develop Fichte's understanding of space as a critique of the transcendental unity of apperception. The Kantian unity of apperception stands as a logical "I," an I whose iteration and synthesis in time gives unity to this purely intellectual entity. Time, it could be argued, stands as a wholly intellectualized account of the process of self-representation. With Fichte's insistence that the "I" be equated with action, that "the intellect . . . be a kind of doing and absolutely nothing more," it is perhaps no surprise that Fichte comes to explain the synthetic power of the imagination, not in terms of time, but as movement, and action (*FW* 1:440).

Imagination and the Synthesis of Space

For Fichte the imagination's original power lies in its capacity to synthesize space or spatial movement. While the imagination has the power to synthesize time, originally all synthesis is of spatial movement. Fichte explains the imagination's role in the synthesizing function of action (or movement in terms of space) in such a way as to explain the emergence of an identity constituted through activity known as the original ego.[15] In the *Wissenschaftslehre Nova Methodo* (1796/99), Fichte begins by explaining sheer acting in terms of the spatial extension of a line.[16] Fichte writes: "The I intuits its 'sheer acting' . . . as an *act of drawing a line,* and hence it intuits its indeterminate power to act in this way *as space*" (*WLnm* 71). Fichte uses the example of a line to demonstrate the synthetic capacity of the imagination to represent movement. The line is an example of our fundamental experience of space as movement.

Kant gives us a further example of how space underlies the notion of time. He makes clear the logic by which space is reduced to time in the process of synthesis. Kant also asks us to think of a line—pure extension. He writes: "When I seek to draw a line in thought . . . obviously the various manifold representations that are involved must be apprehended by me in thought one

15. One recalls that Kant could not achieve such an identity with apperception.
16. Fichte makes use of this Kantian example, but transforms its meaning. For Fichte, it represents the synthesis of space in lieu of time.

after another. But if I were always to drop out of thought the preceding representations (the first part of the line . . .), and did not reproduce them while advancing to those that follow, a completed representation would never be obtained" (*CPR* A102). His point is that the imagination is essential to maintaining the *memory* of previous representations so that at the point of completion the total image can be held in thought. Here then the role of the imagination is to represent "in intuition an object that is not itself present" (*CPR* B151).

Fichte restages this example of the line in the *Wissenschaftslehre Nova Methodo* to explain the dynamic of the imagination, but with a different emphasis. For him the question is that of movement or space. He asks how "movement in the corporeal world is possible" (*WLnm* 398). For him one must begin by overcoming the dilemma posed by Zeno's paradox. According to Fichte, for Zeno, movement is fundamentally *impossible* because in order for a body to move from point X to point Y it must traverse an infinite number of points, and thus can never arrive at its destination. And since one cannot imagine "a movement that traverses infinitely many points," we must conclude that "movement is something that is simply unthinkable" (*WLnm* 398). The paradox according to Fichte is that while movement as such is unthinkable, even a child engages in movement. Fichte offers the following solution to this paradox. He maintains that with movement "one does not think of points at all," but rather "one thinks of {the whole} line" (*WLnm* 398). And it is the power of the imagination that makes the thought of movement possible by giving us the capacity to hold the entire line in thought (memory). The imagination allows us to grasp the infinity of points within a single synthetic act. "The lens" of the imagination allows us to see the event as a whole by means of an image.

The unity of the imagination arises through its capacity to synthesize space. Yet if for Kant self-representation arises in the temporality of the imaginative synthesis, for Fichte the question of the I and its self-awareness is bound up with movement. Kant's exploration of spatial succession in terms of time is for him the only possible explanation, because time itself is the very condition of self-consciousness. According to Kant, time is real to the extent that it is the quintessential mode by which the I comes to appearance in self-consciousness. Kant explains: "Time is therefore to be regarded as real, not indeed as object, but as the mode of the representation of myself as object" (*CPR* A37/B54). Yet while the Kantian ego is the representational ego, the product of time consciousness, the Fichtean ego is a pure act, a self that emerges in the original movement that defines space.[17] As act, the Fichtean ego is precisely anterior to the Kantian understanding of the ego as the product of self-representation (*ego cogito*).

Still, the logic of Fichte's inquiry is clear: he moves from a demonstration of the possibility of movement in general, as a description of space, to an inquiry into the consciousness of our own inner movement or activity as an energy, force, or sheer "doing" (*WLnm* 405). Finally, Fichte clears away any doubts about the significance of space for his new paradigm, by defining it as the central analogy for any description of mind. He writes: "The analogy of movement [through space] comprises the entire function of mind" (*WLnm* 399).[18]

While with the help of the imagination time becomes the schema of the sensible (phenomenal) world that makes representation possible—both the representation of objects and the representation of the I in the transcendental unity of apperception—space, Fichte tells us, is the schema of the supersensible aspect of the self. For Fichte, the schema of the supersensible is acting (*WLnm* 400, 408), while for Kant, of course, the noumenal self is, by his own determination, wholly unknowable.

Through the schematization of time the transcendental unity of apperception stands as a formal I that unites the manifold in a single iterated representation. Thus, for Kant, since we can only know appearances, only the phenomenal representation of the *ego cogito* is given recognition. Yet what of the original ego? According to Fichte, space, and the spatiality of the original active ego, is replaced by Kant through the time-constituted representation of the *ego cogito*.[19] An original unity was given in the spatiality of bodily movement, but the Kantian account demanded a construction.

If for Kant what lies at the basis of sensible representation (as matter in space) is relegated to the sphere of noumena, then for Fichte the schema of the supersensible is acting. What functions as the thing-in-itself—external to sensation—is for Fichte, for all intents and purposes not simply external, but produced. While we suggested earlier that the intellectual aspect of the I and its experience arose in time for Kant (as well as Fichte), Fichte's emphasis on

---

17. Again, this is Fichte's reading of Kant. See Allison, *Kant's Transcendental Idealism*, 286.

18. "On the basis of what has been said above, we have long been acquainted with the principle that states that the I is immediately conscious only of its own activity. Here, however, we asked, 'How is the consciousness of this activity possible?' In order to be able to answer this question, we began with an example: 'we asked how is movement possible?'" (*WLnm* 399).

19. Henry, "Does the Concept of the Soul Mean Anything?" 102.

the role of activity, and power of the body in the determination of space, leads one to wonder whether for Fichte space is not in fact more fundamental than and, in short, a condition of possibility for time and time consciousness. Fichte takes his assertion that "'my body and I'—'my mind and I' mean the same thing" seriously to the extent that he often describes the free intellect itself in terms of absolute space (*WLnm* 321).

Space, Freedom, and the Self-Determination of the Ego

The ideal of absolute space, founded upon bodily movement, grounds the ideal of freedom. Absolute space is the sign of pure determinability. Fichte writes: "Space is empty—in the sense that I—in thought—traverse, empty it and place something therein" (*WLnm* 244). Absolute space stands as the sign of pure determinability for action, thought, and the imagination.[20] Pure space offers us a field of pure determinability and pure freedom. Actual freedom, however, as we shall see, is in the act of determination, in the act of placing an object in space. If we are to understand the role of space in the constitution of consciousness and thus the role of outer sense in the constitution of inner sense, we must better understand this relation between the intellect, freedom, and space.

Fichte could argue that the intellect is not determinable apart from freedom and that freedom is not determinable apart from space. It is the intuition of activity in the form of space that marks the condition of the possibility of freedom. While "agility" (*Agilität*), or acting in general, stands as an indeterminate possibility of freedom, actual freedom arises through the possibility of self-determination—through the positing of a determinate self. But, as we will see, the transition from indeterminacy to self-determination cannot begin without first positing a determinate object, or in this instance, giving an object a determinate place in space.

The act of self-determination through space is the first step in the process of the development of a determinate self or ego. The determination of one's own place in space must be preceded by the determination of the place of an object in space. Fichte explains: "My determination of the place of an object in space must follow from a feeling of myself as occupying a [particular]

20. "When we think of a power, what we think of is the sheer form {of "doing"}, not of any determinate acting of this or that type. This is similar to what happens when we think about infinite space" (*WLnm* 405).

place. Consequently, I must feel myself in space" (*WLnm* 252). It is by means of the determination of the place of an object in space that I determine and become aware of my own situatedness. Space then is not only a form of intuition, it is also a matter of "feeling." This, as we will see, is part of the self-feeling that founds (genetically) one's own self-awareness.

Space is required for the determination of the object, which is in turn required for the very determination of consciousness. Fichte writes: "Space is the subjective condition for the possibility of the object—and space itself is the condition for our reflection upon the object" (*WLnm* 241). Space determines the form of intuition and feeling—for our reflection upon the object and ourselves. Space then, and with it matter (as the extension of space), gives form to representation. Yet like the fantasy of absolute freedom, the indeterminacy of absolute space exists only in the imagination. Freedom, like space, becomes real through determination, and determination is made possible through the activity and movement of the articulated body. Like Feuerbach's critique of God as an anthropomorphic projection, Fichte notes that just as empty space is the abstraction of filled space, so too is the spiritual world an abstraction of the corporeal world (*WLnm* 242). If then there is a spirituality involved in the idea of absolute space (i.e., the notion of absolute freedom), it is Fichte's phenomenology of bodily space—and not Kant's critique (alone)—which truly reins in the excesses of metaphysics.

Spirit, one recalls, is a question of the whole. When something speaks with spirit it gains its voice through an understanding of its relation to the whole. If space is a function of the field of bodily habituation, then like spirit, it gains meaning both within and through its broader horizon of action, which constitutes its field of meaning through those acts. Fichte's recognition of spirituality as an abstraction from the corporeal world (and absolute space, the abstraction of lived space) would allow him, not only to establish the analogy between an abstract spirituality and absolute space, but also to recognize a scale of transition in which the constitution of space as such—in bodily acts—not only takes on a spiritual quality but also exhibits the essential task of spirit. In brief, such analysis must be understood in the context of our greater concern with spirit in the age of materialism. This detailed analysis of space in the generation of consciousness, as a moment anterior to time consciousness, is, as I suggested, a microexplanation (in the register of consciousness) of the stakes of the debate in the discourse of the spirit and the letter. The intimate communicative and representational relation between the spirit and the letter, like the problem of mimesis I have been writing about all along, occurs again in the very generation of the ego and consciousness in the tension between the respective presentational (*Darstellung*) and representational (*Vorstellung*) capacities of space and time. And it should be no surprise that such thoughts will set the stage for our account in the last section of Fichte's theorization of subtle matter as a middle term between the spiritual and corporeal worlds.[21]

Space and Self-Awareness Between the Body and the Imagination

Space, as a feeling, and as a form of intuition, is merely an expression of the dual nature of the I as "striving" and "limitation." Fichte's claim that mind and body are actually perspectives on the same phenomena is again confirmed. If the imagination brought striving and limitation to a synthetic unity through the dialectical image (as we saw in the example of the line), Fichte also described this same dynamic in terms of the power of the body to bring intuition and feeling to a synthetic unity. Where the imagination had given unity to space, and thus to the ego, through the dialectical unity of limitation and striving, it is now the body which gives unity to this phenomena through the dialectical terminology of "feeling" (limitation) and "intuition" (striving) (*WLnm* 254). Before we are able to understand the body, or for that matter the imagination, as a power for synthesizing space, we need to better familiarize ourselves with the antinomic first principles of Fichte's philosophy from which his dialectical concept of the imagination arises. In short, in order to understand the dual sense of space, in the register of the body, as feeling and intuition, and in the register of the imagination, as striving and limitation, we need to explain the first principles of Fichte's philosophy expressed in his well-known formulas: I = I and I ≠ Not I.

As is well known, Fichte believed that first principles in philosophy could not be proved as such, but rather were a matter of choice (*FW* 1:91). For instance, Fichte's understanding of the I as act is an expression of his own choice in first principles. In the *Wissenschaftslehre* of 1794, Fichte expresses his first principle in the following formulas: I = I and I ≠ Not I. For him these formulas express the dual nature of humanity: the human being is a finite being with an infinite vocation. The primordial act of self-positing, I = I, is thus a first principle, which can be neither questioned nor justified,

---

21. We might wonder, for instance, about the space of subtle matter.

but simply accepted at the outset. Here, the I is described as an absolute insofar as the I has achieved identity with itself. As such, it is absolute. Nevertheless, this identity exhibits a separation, even if only in predication (I = I). As a consequence of this separation, the absolute remains an ideal toward which the actual I strives. If the first formula is ideal, the second moment describes the real. This second formula, I ≠ Not I, expresses the finite character of the I known as limitation. As finite, the I is necessarily limited. As experience tells us, the ego is neither absolutely limited nor absolutely free (as a pure identity, the identity of the absolute). Therefore, these two moments must be understood as a dynamic of striving and limitation. According to Fichte the power to express the dialectical character of this dynamic as a unity is known as the imagination.

Our previous descriptions of "action" then must *not* be understood as references to any mere act, but rather to the dynamic of activity and hindrance, striving and limitation, that defines the imagination itself. The original self-positing act of the self is not distinguishable from the first emergence of the imagination. As we noted then, it is no surprise that in his later works Fichte comes to align the act with the image.[22] It is in this sense the act is aligned with the imagination. Fichte describes the imagination in the *Wissenschaftslehre* of 1794 as follows: "The interplay of the self, with itself, whereby it posits itself at once as finite and infinite—an interplay that consists, as it were in self-conflict, and is self-reproducing, in that the self endeavors to unite the irreconcilable, now attempting to receive the infinite forms of the finite; now, baffled, positing it again outside the latter, and in that very moment seeking once more to entertain it under the form of finitude—this is the power of the imagination" (*FW* 1:215; *SK* 193). One only conceives of striving and limitation as distinct moments from a philosophical perspective. Like the drive in which the moments of striving and limitation are inseparable, so too here is each aspect dependent upon the other. Talk of limitation only makes sense in the context of an infinite striving, just as it is the act of limitation itself, which makes possible infinite striving. Striving without resistance would be the static form of pure identity. It is a question of the power of the limit. Fichte explains: "If the self did not bound itself, it would not be infinite" (*FW* 1:214).

The oscillation of the imagination between infinite striving and the finitude of limitation is a movement that allows the self to move from indetermi-

22. Julius Dreschler emphasizes this in *Fichtes Lehre vom Bild.*

nacy to determinability. While the oscillating of the imagination between irreconcilables is a dialectic whose hovering (*schwebende*) movement constitutes time, it is important to recognize that time is generated through movement in space. Time is made possible by a movement that defines space as such. Fichte describes the synthetic power of the imagination to unite the finite and the infinite in terms of the mutual dependency of infinite space and its limitation. For Fichte, space as such would not be possible apart from limitation. He writes: "No infinity, no bounding: no bounding, no infinity; infinity and bounding are united in one and the same synthetic component" (*FW* 1:214). The synthetic power for uniting these antithetical moments of bounding and limitation expresses the function of the imagination and its power as a dialectical image.[23] This process, which gives rise to the image, depends upon a spatial metaphor. The oscillation of the imagination may generate time, but it is time that arises as a function of movement through space. Thus, what the synthetic imagination at base captures through its dialectical image is movement, or in Gilles Deleuze's words "the movement-image."[24] And it is only from the perspective of self-consciousness that the movement-image is describable as time. Still, it should be emphasized that the relation of space to time is developmental. It arises in the development of consciousness. The transition in the form of the imagination's synthetic unity from spatial movement to temporal iteration marks a development in the ego from mere determinability to determinateness. This shift is also what we have been describing as the transition from the original intuition of the ego to its ultimate articulation in self-consciousness.

The determination of the I by means of space is not distinct from our broader conception of the imagination. The imagination stands as a power of determination. Fichte writes: "The imagination is what conditions all activity involved in the act of determining, even though this activity is ascribed to the I alone" (*WLnm* 402). Or again, he explains, "What is determinable in this case does not somehow exist in advance of the power of the imagination; instead, what is determinable comes into being precisely—and only—by means of the imagination" (*WLnm* 402). This determination, the determi-

---

23. It's worth noting that both Benjamin and Fichte conceive the imagination dialectically.

24. Deleuze's description of the theoretical crisis the development of cinema posed to idealist and materialist thought through the notion of the "movement-image," I believe, is quite illuminating in understanding Fichte's own transition from activity to image (in the development from the first to the last period of the Wissenschaftslehre) and the ultimate reciprocity between these two descriptive frameworks, in view of Fichte's own attempt to rethink idealism in the face of a burgeoning technological materialism. See Deleuze, *Cinema I*, 35.

nate I, is a product of the imagination. Fichte describes the entire process by which the imagination comes to give determinateness to the I beginning with its first intuition of mere determinability:

> Beginning with the highest synthesis, one can say, "I intuit myself as imagining, and in this way I observe myself as determinable." {I am not immediately acquainted with myself as something determinate; it is only through the medium of the imagination that I view myself in this way. The imagination serves as a tinted glass through which I view myself, just as, previously, I viewed the object through a tinted glass. The difference is that, in the previous case, the object I viewed through the glass was one that was already there, *whereas in the present case, the object is projected and produced for me at the same time as the glass.}* *To this extent, the imagination is absolutely productive with regard to the content or matter [of consciousness].* (*WLnm* 402; my italics)

Fichte presents us with three clear stages in the progression toward self-conscious reflection. The first stage is that of intellectual intuition in which the self is not yet determined, but merely determinable. The second stage is consciousness of the object. The third stage exhibits full self-reflection in which the I takes itself as object. This reflection takes place through the prism of the productive imagination. Yet just what is the tinted glass of the imagination if not the dialectical unity of striving and limitation, intuition and feeling. Before we turn to understand the body as a power like the imagination for synthesizing and bridging sensible and intelligible worlds, we must begin to understand our relation to ourselves and to our bodies first in terms of the original self-awareness known as self-affection, and then as the kind of knowing known as intellectual intuition. Such an exploration is part of our continued attempt to understand spirit's encounter with technology (a technology of the letter), on the order of the constitution of consciousness, through an exploration of the constitution of space and time.

The "I" as Intuition, Feeling, and the Unconscious

Henri Bergson asserts that "space is not a ground on which real motion is posited; rather it is real motion which posits space beneath itself."[25] Now

25. Bergson, *Matter and Memory,* 217.

Bergson's insight concisely expresses Fichte's point and allows us to refine the loose association we had established between spirit and space.[26] Nevertheless, strictly speaking, spirit, as the "primal sublime," stands beyond all categories of time and space. Spirit was described in terms of space as a means to contrast it with the representational nature of time consciousness. Yet, in reality, as Bergson argues, movement, the vitality of life itself, is that original movement which itself makes space possible. In this case, the immediacy of spirit would be manifest in the notion of activity; its first-order communication would be *presented (dargestellt)* by means of space; a second-order communication would be represented (*vorgestellt*) through time and time consciousness. In other words, like the absolute, spirit as such is unpresentable. It is pure immediacy. The parameters of the problem of presentation or representation before us are still, in essence, the same that defined the paradox of mimesis. In the register of consciousness, time consciousness and self-conscious reflection offer a representational consciousness that was described earlier in the terminology of *Vorstellung*. This presentational immediacy of space—while not activity itself—nevertheless offers a more original *Darstellung*, which, like the spatialization of the plastic arts, offers a liminal form of representation that reveals truth despite its secondary or imitative status vis-à-vis the radical immediacy of spirit (or the primal sublime).

The act is original. The act is the vital immediacy one aligns with spirit. The question is how one is aware of that act. How can one be aware of that activity, simultaneously with that activity, without thereby condemning that vitality to death in the act of self-conscious reflection and representation? To answer such a question we need to chart the origination of the ego from its inception to a full-blown self-consciousness. In particular, since Fichte's version of intellectual intuition is traditionally understood as a form of immediacy (despite its complicity with time consciousness), I will be especially concerned to differentiate it from that still more immediate form of self-awareness called self-feeling.

It is the act, the action of striving and limitation embodied in that bodily movement, which constitutes space. It is the articulated body through which one's own spatiality is constituted. And it is through this spatiality that the original ego and self-awareness first arise. Just as the imagination is the syn-

26. Bergson's statement also makes clear the importance of Fichte's work for the phenomenology of the body. One could, for example, trace Fichte's work to more contemporary phenomenologies of the body, Merleau-Ponty's for example, through the ideologue tradition of Maine de Biran. Biran was greatly influenced by Fichte, and he, in turn, greatly influenced Merleau-Ponty.

thetic unity of striving and limitation, the body stands as the synthetic unity of intuition and feeling. The body then, as an articulated body, is a system of self-feeling, an original immediacy to itself that anticipates the act of reflection.[27] Fichte's exploration of the condition of the possibility of the Kantian *ego cogito* led him back from the culmination of self-consciousness—in representational consciousness—to that more immediate form of awareness made possible in self-feeling and the spatiality of bodily movement. For Fichte self-consciousness was not an original moment, but the final moment in an ever more cognitive or intellectually oriented self-awareness. According to Fichte, the progression toward self-consciousness proceeds as follows: limitation, feeling, intuition, consciousness (representation of the object), and finally self-consciousness (representation of the self as object). We now turn to articulate this path of development in more detail.

From Limitation to Feeling

The first moments of this progression, in the transition from limitation (*Begrenzung*) to feeling (*Gefühl*), begin with the body. We noted that Fichte's emphasis on space and the active body allows him to subvert the purely representational account of consciousness by establishing the very conditions of possibility of time-constituted consciousness by delimiting the concept of time itself in terms of spatial movement. Thus the first emergence of the self arises out of the dialectical determination of space in terms of striving and limitation. Space, however, is not divorced from the problem of feeling. In other words, the very problem infinite freedom poses for a finite individual first appears in terms of the infinity of space, a space whose ultimate determination is one with the articulation of the conscious self. The intuition of infinite space offers the intuition of infinite striving; but since, as we noted, "no limitation, no striving," one pole will require its dialectical opposite (limitation). Here striving would be meaningless without consciousness, and consciousness requires limitation. Consciousness does not begin with us, but with the intuition of objects. In other words, "only in the intuition of an object do I first become an intuiting subject and comprehend myself" (*WLnm* 255). One finds oneself in space through the simultaneous intuition of an object in space.

---

27. For Fichte's discussion of the "articulated body," see *WLnm* 459.

The difference between limitation and feeling is significant. Limitation is resistance to striving. Limitation merely "is," while feeling is an awareness of that limitation. Feeling is a form of affection: it is self-affection (auto-affection).[28] Feeling is that original form of self-awareness which arises in that essential dyad or dialectic of the drive and its limitation. Thus Fichte can claim that feeling is what marks the limit between life and death (*FW* 1:299). It is the original moment. The first self then, constituted in feeling, emerges as an original spontaneity in the materially dependent dialectic of striving and limitation. For Fichte this combination of striving and limitation, known as the drive, forms the basis of original reflection. Here the self is identified with the drive. The drive constitutes original reflection: "no drive, no reflection" (*FW* 1:294). Yet what sort of reflection is this? In what sense is this original reflection prior to self-conscious reflection (in which the subject takes itself as object) actually a form of self-reflection? The first order of reflection is not reflective in the speculative sense: it is a question of feeling, of self-affection. Here the subject is *not* its own object. Rather, in feeling, there is an innate immediacy.[29] In this activity the self is both "what feels and what is felt, and therefore stands in reciprocity with itself" (*FW* 1:300).[30] The body does not simply take itself as object; it is simultaneously subject and object, feeling and what is felt.[31]

The drive constitutes reflection to the extent that striving is limited and thrown back upon the self. It is this impediment or limitation to striving which provides the resistance that makes feeling possible. The force of striving that is turned back upon the self in the act of limitation marks the first

28. My reading of Fichte's attempt to articulate an original self, anterior to the dualism of self-reflection, in the immediacy of self-feeling, is greatly inspired by the work of the phenomenologist Michel Henry. See, for instance, his *Genealogy of Psychoanalysis*.

29. As Rudolph Gasché explains the dilemma, the paradox of the philosophical concept of the dualism of self-conscious reflection is potentially resolvable through either absolute reflection (i.e., Hegel), or through the notion of an immediacy. Both, as Gasché argues, are problematic. In particular, he denies the validity or possibility of immediacy. I, however, disagree with him on this latter point. See Gasché, *Tain of the Mirror*.

30. Such claims anticipate Merleau-Ponty's subjective body and his critique of reflection.

31. While Dieter Henrich had spoken of the original self, and Fichte, "self-feeling" (*Selbstgefühl*), it is Michel Henry who makes explicit use of the term "auto-affection," and as a consequence, makes clear—although without always explicitly engaging the work of Fichte—the relation between the constitution of an original self in auto-affection, and its ground in the social phenomenon of affective rapport. In other words, in the same way that Fichte had argued in intersubjective recognition that the recognition by the other was necessary for self-consciousness and freedom, I am arguing that Fichte's search for the ground of recognition and self-consciousness in rapport would lead him to recognize—much on the model of intersubjectivity—that self-feeling, or auto-affection, was constituted in hetero-affection, in and through an affective influence of the other.

expression of reflection and the representational drive (*Vorstellungstreib*). This primordial structure of representation is not on the order of self-consciousness proper—of what would come to define a true reflection in which the subject takes itself as an object. At this moment the self is merely felt. Fichte writes: "Since in this reflection the self is not conscious of itself, *the reflection in question is a mere feeling*" (*FW* 1:297; my italics). Although it is not yet conscious, the self at this stage is still posited as a self (*FW* 1:298–99). It is this original self, as we will see, which Michel Henry, following the work of Maine de Biran, would define explicitly in terms of the unconscious.[32] Like the I as noumenon we referred to earlier, here too the self appears prior to consciousness, and is in this sense unconscious.

This *unbewußt* aspect of the body and bodily movement is essential in the development of the synthetic power of the imagination. The dialectic of striving and limitation finds unity in both the physical register of the body and the intellectual register of the imagination. Through original activity (striving) what is felt is oneself by virtue of a resistance to one's infinite movement. In more typical Fichtean language, the "I" of striving comes up against the Not-I of limitation—the *Anstoß*. These terms, however, remain mutually dependent. But how does one understand this dependency? Fichte's circular description gives us a clue. Like the fantasy of the Kantian thing-in-itself which appears under the condition of its inaccessibility, so too does resistance to action make infinity itself possible. The freedom associated with infinite striving and the limitation brought on by the *Anstoß* is itself a self-evident fact of consciousness that cannot be demonstrated or readily defined. After all, as Fichte writes the following in the first sentence of the *Wissenschafts-lehre* of 1794, just before he expresses the idea of infinity (A = A) and limitation (A ≠ Not-A) in logical form: "Our task is to discover the primordial, absolutely unconditioned first principle of all human knowledge. This can be neither proved nor defined" (*FW* 1:91). Yet these assertions from the theoretical part of the *Wissenschaftslehre* of 1794 not only seems to bear themselves out in the practical portion, "it is this practical part alone that speaks of an original reality" (*FW* 1:285–86).

While infinite striving and limitation were posited as necessary theoretical assumptions in the first part of the *Wissenschaftslehre* of 1794, it is in the second part, in the practical portion, that Fichte shows how this dynamic

---

32. Although Henry's task is more to distance himself from the conscious/unconscious opposition.

mutual dependency first appears in the development of consciousness. Since feeling is the very first moment of any experience whatsoever, its reflective vocation is inherently one of self-feeling (*Selbstgefühl*). Feeling is the content and effect of the formal aspect of the *Anstoß*. Yet while feeling arises in limitation, following Fichte's insistence on the mutual dependency of limitation and striving, we quickly recognize that this reflective awareness sets in motion the self-perpetuating dynamic of limitation and striving by which the intellect constitutes for itself an objective world.[33] Thus feeling expresses the unity or synthesis of self-striving and self-limitation, thereby marking the first appearance of the self as drive and as primordial reflection. As a result, "what feels is therefore posited as a self" (*FW* 1:299).

Here self-limitation, as feeling, functions as the basic structure of self-representation (*Selbstdarstellung*). For Fichte, feeling defines the self and an original reflection, to the extent that it is at once "what feels and what is felt, and stands in reciprocity with itself" (*FW* 1:299).[34] Despite the language of reflection, at this stage, the self's relation to itself does not take itself as an object as with self-reflection proper. Here, if there is a schism, it is not yet a subject/object dualism, but merely a first rupture, a hairline crack or break within nature that, as with our earlier discussion with mimesis, would require merely a mimetic identification in order to bridge that gap. Yet, even in its self-immediacy, it is a gap nevertheless, which is the first moment in the development of self-consciousness.[35]

From Feeling to Intellectual Intuition

Just as "feeling" marked the boundary between life and death, the transition to intellectual intuition marks the limit between life and intelligence (*FW*

33. Daniel Breazeale emphasizes this point in his article "Check or Checkmate." See "Check or Checkmate," 95.

34. In fact, it is the very basis of cognition: "All of our cognition does indeed begin with an affection, but not with affection by an object" (*SK* 74).

35. Such a view puts me between two positions. For instance, while thinkers like Rudolph Gasché vehemently deny that one can overcome the difficulties of self-conscious reflection through a recourse to immediacy, others like Michel Henry offer compelling solutions precisely through descriptions of immediacy in the discourse of auto-affection. By virtue of our discussion of mimesis, and mimetic identification in the break from nature, however, my position, while leaning toward Henry's, nevertheless recognizes at least some break, however minimal, has occurred. Consequently, any discourse of immediacy must take such a break into consideration. Unlike Gasché, however, I do not believe that a rupture commits one to the discourse of reflection. Rather, as I suggested,

1:298). What is felt in the act of self-feeling is the transition from the "feel-ing" of limitation to the "intuition" of oneself. This is a transition from grasping feeling as one's own to the act of intuition understood as feeling "feeling" as one's own (*WLnm* 192). What the I intuits in this limitation is the feeling of its self-active self. Here, however, true self-consciousness has not yet been achieved. In intuition I am only the subject of intuition, not its object (*WLnm* 194).

This original feeling of the self prior to consciousness and a full-fledged self-reflection sounds strikingly like what is described in the "Second Intro-duction to the Wissenschaftslehre" as "intellectual intuition." Of course for Fichte, intellectual intuition is *not* that sort of immediate knowing of the Thing-in-itself which Kant so vehemently critiqued, but rather something quite different (*FW* 1:471–72). Fichte writes: "Intellectual intuition is the name I give to the act required of the philosopher: an act of intuiting himself while simultaneously performing the act by means of which the I originates for him. Intellectual intuition is the immediate consciousness that I act and of what I do when I act. It is because of this that it is possible for me to know something because I do it" (*FW* 1:463). Intellectual intuition then is the intuition of oneself as a "sheer activity." Thus for Fichte this original act is a fact of consciousness or, in his words, a *Tathandlung* (*FW* 1:465), and as such, defines any true philosophy (*FW* 1:466). Yet while intellectual intuition of this original self-activity is said to mark an original consciousness (con-sciousness of the I comes only from intellectual intuition), it is not clear it could be identified with feeling, since feeling is said to precede consciousness.

Fichte explores the very limit of consciousness. If what defines that limit or boundary are those sides which lie within and without it—thereby marking it—then feeling can be said to express something unconscious or prior to consciousness which defines its very condition of possibility, while intellec-tual intuition is that first instance which signals conscious self-awareness. In the *Wissenschaftslehre Nova Methodo* Fichte asks himself the same difficult question: "What then is the nature of the transition from feeling to intu-ition?" (*WLnm* 180). He responds: "I cannot intuit a feeling unless it lies within me; thus if I am to intuit a feeling, I certainly have to be a feeling

---

through reference to Adorno in an earlier chapter, such a minimal break is what inaugurates mimesis and a mimetic identification with a nature one has just broken with. Consequently, although the very discourse of rapport, and affection, like mimesis, requires some minimal break, it is my view that such a break does not yet commit one to a full-blown theory of reflection. See Gasché's *Tain of the Mirror*.

subject. An act of reflection simply occurs [at this point]. By means of a new act of reflection, . . . the I, as intuiting subject {looks down upon the latter as substrate}, and thereby becomes independent" (*WLnm* 181). The movement from self-feeling to self-intuition marks the true arrival of consciousness, and this transition, at least according to the *Wissenschaftslehre* of 1794 can only be described in terms of a leap (*FW* 1:298).[36]

Intellectual Intuition and the Materiality of the Soul

As Pascal well knew, one leaps when faced with an otherwise unbridgeable gap, a gap or transition that cannot be rationally justified. It marks the very limit of reason. Nevertheless, if Fichte's own narrative is plagued by this gap—and consequently, the required metaphysical leap—it is precisely this transition, the transition between sensible and intelligible worlds, which haunts him. Like the problem of the manifestation of spirit through the letter, here we are confronted with the translation of an original, spatially constituted self-awareness of the ego into the order of time. Space is translated into time. That which is intellectual or spiritual must be sensibilized (*versinnlicht*). Our concern with such an issue should not be surprising since our guiding problem is the materialization of the imagination, which culminates in Fichte's work with his encounter with magnetic psychology.

In the imagination it is time which brings the ineffable to intelligibility. In Fichte's words, it is time which brings the noumena to actual intelligibility.[37] The paradox of such an assertion is that noumena—one imagines—stand precisely beyond time. But what Fichte is demanding, again, is a leap. A leap, whose very gap he was to wrestle with for the next twenty years in various forms, from the notion of intellectual intuition to his account of subtle matter in the *Grundlage des Naturrechts* to his last description of magnetic fluid and the materialization of spirit in terms of a "Physicirung des Idealismus" in his last works (1813). It should be clear then that the chasm which demands this leap is not distinguished from our long-standing discussion of the problem of representation understood as mimesis.

Fichte's critique of representational time consciousness obviously did not

36. "From this absolute spontaneity alone there arises consciousness of a self . . . not through transition, but by means of a leap" (*FW* 1:298).

37. He writes: "Noumena is what is intelligible therefore, made sensible by means of time" (*WLnm* 282).

relegate the concept of time to insignificance. In fact, Fichte writes: "The concept of time is of decisive importance for the system of the Wissenschafts-lehre, as it is for idealism generally." He continues: "Everything depends, above all else, upon the genesis of the concept; because only in this way [that is by indicating how the concept of time is generated] can critical idealism be justified" (*WLnm* 277). The concept of time is crucial to critical idealism because without it, without the unification of the manifold through time, the self-conscious transcendental subject could not be given unity and stability. Yet if one reads Fichte's language closely, one sees that he does not simply assume time, but rather he seeks its genesis, its conditions of possibility. The difficulty in unraveling this genesis, however, resides in the problem of our original limitation or finitude.

Fichte tells us that while Kant posits two orders of time, the subjective and the objective, in which one is subsumed by the other (and Reinhold posits two different spaces), he in fact intends to avoid such divisions. For Fichte our finitude presents us with certain roadblocks to questions in which one cannot, for instance, infer an absolute time. Rather time, like space, is expressed for Fichte phenomenologically. He writes: "{. . . I cannot feel what is feelable 'as such' or 'in general,' but only in part. I feel what is rough only as something rough, not as something feelable in general}. This is the limita-tion I encounter in grasping my own state" (*WLnm* 278). Fichte insists that "this limitation makes and constitutes the limit of reason" because "reason comes into being only by means of this very limitation" (*WLnm* 288). In other words, "so long as one expects a rational answer, one cannot ask any questions concerning this limitation" (*WLnm* 278). "Rational" in this con-text means "thinking," discursive thinking made possible by the order of time. Yet if time and discursive thinking arise contemporaneously, as Fichte points out, and discursive thinking is associated with rationality, then Fichte's inquiry into the origin and generation of time precisely exceeds the frame-work of rationality.[38] In this sense—as I have now emphasized several times—space, or at least activity, defines the realm of the *Unbewußt*.

Time stands as an index of change. Yet while "one must think of several united moments, the question remains: united in what?" (*WLnm* 280). Fichte reproached Kant for instituting two forms of time: the subjective and objective. Yet Fichte's description of intellectual intuition seems to reinscribe the same distinction. Intellectual intuition, as the very form of time, appears

---

38. A point that would be no surprise to contemporaries like Schelling.

to stand both within and without the form of time. Thus intellectual intu-ition is said to be on the one hand "repeated over and over again throughout all discursive thinking and is thought of in every moment as the same," and on the other hand, "time is the form of intellectual intuition" (*WLnm* 280–81). In other words, intellectual intuition stands as the key to the paradox of time which requires the simultaneity of alteration and stasis. There must be change, but change in the face of an unchanging observer. The self is split: "I observe myself within time, but I am not in time, insofar as I intuit myself *intellectualiter* . . ." (*WLnm* 280). Fichte's point is that, at the limit of rationality, intellectual intuition stands both within and beyond time. It is the single act of willing, the original intuition of the self, which gives unity to change. The self of intellectual intuition stands to account for both tempo-ral change and the static unity in which time inheres, thereby giving it mea-sure. Here the I is not the logical affirmation which Kant concludes must accompany all representations in his description of the transcendental unity of apperception. Rather, its unity is original. The self precedes all representa-tion (*Vorstellung, Nachahmung*). Fichte, in a clear critique of Kant, writes: "Unity must already be present before I can combine anything" (*WLnm* 280). Or, in other words, there is an original I anterior to any synthesis.

Intuition offers the original unity that makes time as such possible, and as a consequence, the temporal intuition that allows one to infer this original unity as an intellectual intuition. The will, or intellectual intuition, however, is nevertheless sensibilized through time. It is what makes the self discursively real. As a consequence Fichte is able to claim "time is the mediating link between what is intelligible and what is sensible" (*WLnm* 281). If, as we noted, Fichte identified the original self with the noumena and "noumena are what is intelligible, insofar as this is made sensible by means of time" (*WLnm* 282), then Fichte's entire inquiry into the foundation of time and time consciousness can be understood—according to his own argument—as an attempt to articulate (by means of time) what exceeds the boundary of rational inquiry (like Kant's postulation of noumena). But what is the self beyond rational inquiry if not the soul?

Self-Feeling and the Soul

In the context of the ideal of communication (defined by a pure exchange between spiritual beings) the attempt to reformulate and articulate the concept of the soul has a newfound importance. The notion of the soul, as spiritual substance, has been out of fashion at least since Kant. In the First Critique Kant's reining in of metaphysical speculation would, in effect, discard the metaphysical notion of the soul. While in the rational psychology Kant purports to give us real knowledge of the soul, it is actually an exercise in pure thought that develops the transcendental predicates of substantiality, simplicity, identity, and so on. This exercise in pure thought cannot be a true knowledge because it is not a synthetic knowledge: a logical ego is established but is accorded no real being.[39] Nevertheless, in view of Fichte's critique of Kant's formal ego, and his own description of the ego in terms of a prereflective noumenal force, one wonders whether Fichte himself has not returned to a metaphysics of the soul.

Even though, in his description of the noumenal ego, Fichte articulated the soul, he had not, one can be sure, returned to the idea of soul as substance.[40] After all, the ego is defined as act. Fichte is quite clear on this point: "The I is not the soul which is a type of substance" (*WLnm* 112). More positively, in the "First Introduction to the Wissenschaftslehre" Fichte describes the soul not as a thing, but as the product of interaction, as the effect of striving and limitation. He writes: "The soul must not be any special sort of thing at all; it must be nothing whatsoever but a product of interaction between things" (*IWL* 23). The soul then is not substance, but—if one is to speak of it—it should be understood as the product or effect of action and interaction.

Fichte's description of the soul becomes still clearer. In the *Wissenschaftslehre Nova Methodo* he explains that the soul is the organ of inner sense, and the body, the organ of outer sense, and that the soul and the body are merely two aspects of the same original ego (*WLnm* 341). The original power of the noumenal ego then, as duration, iterated in the schematization of time, appears as the soul; and sensibilized through outer sense, extended in space, it appears as the body (*WLnm* 458). Consequently, soul and body stand as products of two respective forms of intuition—inner intuition and outer intuition—two intuitions of the same original ego.[41]

Fichte's description of the soul and body as merely aspects of the same power opens the way for various materialist psychologies. For instance, when

39. Michel Henry, "Does the Concept of the Soul Mean Anything?" 95–97.

40. Kant understands the soul, like other noumenal entities, to be a *"substantia noumena"*—between substance and materiality (*Materialität*). Allison, *Kant's Transcendental Idealism*, 286.

41. With Fichte though, it is outer action that gives rise to inner sense.

Fichte writes, "One treats the body and soul as one and the same: e.g., 'I cut myself' and 'I think I have cut myself'" (*WLnm* 420), he suggests that actions upon the body are actions upon the soul and vice-versa. Further, if the soul can be approached through the material forms of the body, then our knowledge of the soul need not be relegated to mere speculation.

Insofar as the soul, as an activity, has the characteristics of a material body, but is not material, and thus, in short, stands between material and immaterial descriptions, it might be described by the term "materiality" (*Materialität*). Materiality stands as a limit concept between material and immaterial worlds. The notion of materiality attempts to designate that elusive middle term between spirit and matter. The conception of the materiality of the soul (*die Materialität der Seele*) was not Fichte's idea alone. It was already being explicitly expounded by many like Michel Hissmann, in his work of 1777 *Der Psychologische Versuch.*[42] For Hissmann the materiality of the soul simply designated "feeling the thinking capacity of the bodily organ."[43] As a consequence the concept of the soul could be brought into relief through experiment and observation. What still stood as a problem, however, was the translation of spiritual into material forms, and the material into the spiritual. What the notion of the materiality of the soul offered, however, was a way to think a middle term between the rigid binaries of spirit and matter. Yet while Fichte's own attempt to move away from the notion of the soul as substance by articulating the original ego in terms of activity gives some distance in thinking the soul as a middle term, it is not until Fichte's description of "subtle matter" in the *Grundlage des Naturrechts* that he confronts the possibility of a spirit to spirit communication by means of a third term.

Self-Feeling Between the Soul and the Cogito

Fichte's description of an original self-affective self seems to put an end to the infinite regress of representation. It is no longer that parody of idealism that claims it is "representation all the way down." Still, his reformulation of the *cogito* has not been adequately assessed. A number of questions still remain. For instance, what does the *cogito* have to do with our broader discussion of spirit, and how is this inquiry into the self-representational power of the "I" an issue of communication?

42. Cited in Finger's *Von Der Materialität der Seele,* 45–46.
43. Ibid., 46.

To begin with, the I and its self-relation stands as the original moment of communication. And the tension between the spirit and the letter stands as a paradigm of that I's communicative exchange in which the letter designates the sphere of representation (*Vorstellung, Nachahmung*), space, presentation (*Darstellung*), and spirit (the realm of affective immediacy). Like the banality of the purely mechanical exchange of the letter, the representational self for Fichte cannot stand as the source and communicative ground of spirit. Rather, it is the self-affective self, the self of "life" beyond representation, that stands as the source of spirit.

What links both self and spirit is the key term of communication and community known as affection. As we will see in the next chapter in our discussion of subtle matter, the fundamental bond of community has its source in the I's first emergence in self-affection. This line of thinking, however, offers an account of the social through an affectivity prior to the transcendental deduction of right (as outlined in the *Grundlage des Naturrechts*); however, we need to detail precisely in what sense the paradigm of affectivity presented by Fichte's original self threatens the transcendental account of subjectivity.

We begin by returning to Descartes, the Descartes as interpreted phenomenologically by Michel Henry—which, as will become clear, is a Descartes much in the spirit of Fichte.[44] In the Second Meditation, after his well-known invocation of the evil demon, Descartes's hyperbolic skepticism casts everything into a radical doubt—including his own vision. Yet still, there is certitude. Descartes exclaims: "At least it is very certain that it seems to me that I see light, hear noise and feel heat."[45] What then is the source of this certitude? *"At Certe Videre Videor."*[46] This certitude is not on the order of visibility, of *videre,* because it must remain invisible and thus anterior to the illusions of the evil demon. Rather, as Michel Henry suggest, certitude resides in the register of *videor,* in the productive "capacity of original appearance and gives itself in the truth of which vision manifests itself and gives itself originally to us."[47] Here, the certitude of the cogito arises not in the reflexive act of representation in which I see myself seeing, but in a seeing that is a primitive self-awareness beyond the self-knowledge defined in terms of self-

44. One might even see this reading to have its origins in Husserl's reading of Descartes. See, for instance, among other works, his *Cartesian Meditations.*
45. Cited in Henry, *Genealogy of Psychoanalysis,* 17–18.
46. Ibid., 26.
47. Ibid., 22.

consciousness. Following his assertion "it seems to me that I see," Descartes further pinpoints the certitude of the *cogito* by describing the primitive seeing or self-awareness as follows: "This is properly in me what is called feeling and thus, taken in the precise sense, is nothing other than thinking."[48] Here the *cogito,* that self-feeling, auto-affective self, appears with certitude as a thinking that is profoundly anterior to discursive knowledge and consciousness. And it is in this sense that the original self is not opposed to the unconscious, but calls into question very conscious/unconscious binary itself. It is simply *das Unbewußte.*

For Henry the *cogito* is life: it is itself pure phenomenological light.[49] The *cogito* is that invisible light which makes vision possible. In Henry's words, "A thing that thinks is nothing other than the flash of lightning . . . the materiality of phenomenality as such."[50] For Henry, the beyond of all transcendence is the beyond of a radical immanence that is itself that power of being with which everything starts: life.

Fichte too confronted the paradox of reflection by articulating a light whose radiant immanence is not equatable with mere phenomenal visibility. Building on the immediate awareness associated with auto-affection and intellectual intuition, Fichte (like Henry's interpretation of Descartes) turns to describe the immediate self-awareness beyond conscious knowing within the language of vision, at an intersection between the phenomenal radiance of the original light of life and its material form of affection. Thus, as we will see, this original source of self-awareness, the primal knowing of knowing, ultimately becomes transformed in the development of Fichte's work, in the *Wissenschaftslehre* of 1813, into that metaphor of invisible light, the seeing of seeing (*das Sehen des Sehens*).

Like Plato, true knowledge for Fichte will ultimately be identified with a "unity and light" that is "qualitatively absolute" and "can in no way be grasped conceptually" (*FW* 10:259). The problem of self-reflection for Fichte literally becomes the problem of the seeing of seeing. The transition within Fichte's work becomes apparent when one recognizes that the terminology of the seeing of seeing in his last works may merely bring to relief that mute phenomenal source present (earlier) as the self-active self, whose original self-

48. Ibid.

49. Ibid., 7.

50. Ibid., 20. Yet what is a material phenomenal light—subtle matter? One might also look to Fichte's own references to light and subtle matter in the *Grundlage des Naturrechts.*

awareness in self-affection was but an expression of that triumphant radiance by which our acts are to be but vehicles of the Absolute.

Thus, Fichte affirms that while we cannot see the eye itself, "what the eye sees is the seeing of the eye" (*WLnm* 129). Like the thinking of thinking that reveals a *reines Denken* (pure thinking), the seeing of seeing nevertheless produces a certain transparency, productivity, or original Being which shines through. Thus, this discourse of the eye anticipates Fichte's later claim that the I is fundamentally *durchsichtig* (transparent).[51] And transparency offers a *radical transcendence through immanence* in which this original self-affection is simultaneously a self-transparency that is a giving of oneself over to the Absolute. While we are at this point clearly anticipating what is to come, such an overview makes clear why Fichte would make the otherwise rather curious assertion that "freedom is absolute self-affection, and nothing more" (*WLnm* 159).[52] Both freedom and immediate self-consciousness stand beyond the discursivity of intuition. It should be no surprise that Fichte, whose claim that his philosophy was fundamentally a philosophy of freedom, would then become so involved in the articulation of that absolute self-affection that—according to his claim above—he would define freedom in terms of self-affection. If freedom stands as the ultimate achievement of this dynamic of self-affection, then surely Fichte's account of the social, even as early as the *Grundlage des Naturrechts,* cannot remain blind to this trajectory whose seeds were sown—I am arguing—as a critique of reflection as early as the *Wissenschaftslehre* of 1794.

---

51. See the *Wissenschaftslehre* of 1813 in *FW,* vol. 11.

52. My italics. Or again, "Formal freedom is immediate self-consciousness itself, and is therefore unintuitable" (*WLnm* 159).

5

Subtle Matter and the Ground of Intersubjectivity

Introduction to the Transcendental and the Empirical

This book began with the historical observation that the transcendental account of the social sphere (as defined by the transcendental account of intersubjectivity and the intersubjective recognition of right) was profoundly transformed and threatened by a growing predominance of the technology of reproductive media. Reproductive media, I argued, transformed the very medium of human interaction in such a way that the transcendental account no longer seemed to describe one's actual, everyday experience of it.[1]

Recognizing that, for Fichte, the self/other relation was largely constituted through the unifying power of the imagination, I suggested that what repro-

1. I am not suggesting, of course, that Fichte's account of intersubjective right is no longer of value, but rather that the ever-growing predominance of reproductive media in our lives demands a new account of human interaction and recognition.

ductive media's technically produced power of imaging threatened for the intersubjective account was precisely a crisis of the transcendental imagination. While in the remaining chapters I will argue that the transcendental imagination undergoes a double displacement, thus far we have established the background for such claims by exploring the imagination in terms of a medium which Fichte hoped could somehow mediate between spiritual and material, real and ideal, worlds. What Fichte sought in the synthetic power of the imagination was something like a third term that could partake of, and thus synthesize and mediate, each realm. What he sought was a medium that was not merely representative (in the sense of *Vorstellung*), but was itself immediate. As we saw in the previous chapter, Fichte conceived the synthetic power of the imagination to arise through activity, through movement in space, and, in short, through the immediacy of the self-active ago. Yet how are we to understand Fichte's exploration of immediacy if he affirms that "all transcendental experience comes to end with immediate feeling" (*IWL* 75)?

While one might read this statement from the "First Introduction to the Wissenschaftslehre" to be merely part of Fichte's attempt to define the scope or limits of transcendental experience, such a reading would fail to take into consideration the developmental character of Fichte's lifelong project of the Wissenschaftslehre and the power of the synthetic imagination to elevate human thought beyond apparent binaries. What I am suggesting is that Fichte's shift from an account of transcendental subjectivity in the first period of the Wissenschaftslehre to a metaphysics in the last period had its roots in Fichte's own sense that if the imagination was to synthesize the dual series of the real and the ideal, and thus move us beyond the limitations of each, then certainly this could not be presented from the perspective of transcendental subjectivity alone.[2]

While it is widely recognized that Fichte ultimately moved beyond the transcendental account of the imagination in the last period of the Wissenschaftslehre, the details of such a development are rarely confronted.[3] As part of my attempt to explain the role of the notion of subtle matter in Fichte's philosophy, I would like to suggest that (as odd as it may sound) *subtle matter stands as something like an empirical or quasi-material instantiation of the transcendental imagination.*[4] In order to develop this clue in the relation between

2. This is the exact problem Schelling discusses in *The System of Transcendental Idealism.*
3. See Dreschler, *Fichtes Lehre vom Bild,* and Janke, *Fichte: Sein und Reflexion.*
4. Such a claim will become clearer in our discussion of Fichte's notion of the real and ideal series.

subtle matter and the imagination, however, we need to better understand Fichte's own conception of the role the empirical plays in his understanding of transcendental philosophy. And from here we will come to explore Fichte's very first references to the discourse of subtle matter in the *Grundlage des Naturrechts*. What needs to be explained is why Fichte would make reference to this rival, affective account of the social in what is perhaps his most important theoretical contribution to social theory: his transcendental deduction of right. In order to understand why Fichte would reference subtle matter in his transcendental deduction of the other, we need to begin to understand the role Fichte accords the empirical in his reformulation of Kantian transcendental philosophy.

At first glance, Fichte's description of the social in the *Grundlage des Naturrechts* seems indisputably transcendental in nature. After all, he deduces the necessity of the other as the condition of possibility of his own freedom and self-consciousness. Yet is Fichte's account exclusively transcendental in nature, or does he transform the transcendental in such a way the he must also take into consideration not merely a hypothetical, concrete other, but an actual, material other?

For this project such a question arises out of an attempt to explain Fichte's references to magnetic psychology's discourse of "subtle matter" in the *Grundlage des Naturrechts*. Fichte employs the language of subtle matter as a third term that would mediate between intellectual and material influence. What he is attempting to explain, for instance, is how the intellectual influence of recognition can cause the material effect of bodily restraint. While the details of such an account are for the moment unnecessary, the question remains: Why would Fichte engage in attempting to explain the *actual dynamic* of intersubjective recognition (*Anerkennung*) and the summons (*Aufforderung*), if his intent was merely to deduce the other as transcendentally necessary for freedom and self-consciousness? Why look for an empirical verification of a claim that is exclusively transcendental?

Not surprisingly, the answer to such a question is complex. I will begin by arguing that Fichte developed and expanded Kantian transcendental philosophy in such a way that transcendental philosophy could no longer remain wholly independent of the historical and empirical world. By both developing it as a process and recognizing that the transcendental account could no longer be understood to be ontologically constitutive of the world, Fichte ultimately loosened the transcendental structure in such a way that it required a "pragmatic verification" or perhaps, more precisely, an empirical falsifiability in the historical lifeworld.[5]

Such an account is already inherent in the structure of the project of the Wissenschaftslehre. Fichte concludes, in the well-known account in the "First Introduction to the Wissenschaftslehre," that since first principles cannot be proven, the initial impasse between idealism and dogmatism can only be decided on practical grounds. This conclusion is a victory for idealism precisely through such a pragmatic appeal: Fichte sides with freedom. Despite this conclusion, however, such a decision can never eradicate the specter of dogmatism, which Fichte identifies with empiricism, realism, and materialism (*IWL* 15–17). In other words, Fichte's dismissal of dogmatism in favor of idealism does not discount the possible truth of dogmatism, but merely discounts such a choice from practical considerations alone. In fact, the fundamental task of the Wissenschaftslehre is to establish an ultimate unity or synthesis of real and ideal worlds. This is the task for that power of synthesis called the imagination.

Now what one should start to see is how Fichte's philosophical concern with the imagination and its synthetic power to unite real and ideal worlds intersects easily with the late eighteenth-century discourse of "subtle matter" and its aspirations to unite material and immaterial worlds. While I am not claiming an identity exists between these still diverse explanatory approaches, it is a historical fact that Fichte saw the concerns of subtle matter and its discourse to be similar enough to his own philosophical concerns that he not only sought to study it extensively, but also hoped to integrate it and make use of it in his own philosophical work.

---

5. Robert R. Williams uses the language of "pragmatic verification" and "post-transcendental verification" in *Recognition*, 37–43. Since my first draft of this work I have argued elsewhere that Fichte does not use the empirical directly to "verify" transcendental claims. Rather, I have argued that, like philosophers of science like Karl Popper, Fichte attempts to use the empirical not to verify but to falsify transcendental claims. See my article "Falsification: On the Role of the Empirical in J. G. Fichte's Transcendental Method." Yet despite my argument in this article and the substantial textual evidence that supports it, I am more and more convinced that Fichte seems to want it both ways. On the one hand, he acknowledges the value of the empirical for falsifying transcendental experience and thus upholding the rigor of the Wissenschaftslehre as *Wissenschaft*. On the other hand, his need for his transcendental philosophy to be true is so strong, he seems to want far more than that negative sort of verification which falsification provides. In other words, despite his explicit recognition of the need for the empirical to operate negatively as a falsification only, for any true *Wissenschaft*, he nevertheless continues to aspire for a verification in the positive sense—and his unrelenting search for a physical proof for idealism, for a "Physiciring des Idealismus," would seem to bear this out. As a consequence, the waffling between the terminology of verification and falsification is not interpretive sloppiness or an inattention to language on my part, but a reflection of Fichte's own indecisiveness.

Fichte is widely regarded to have dispensed with the transcendental subject in the last period of the Wissenschaftslehre in his transition of emphasis from subjectivity to being.[6] This shift from a largely transcendentally inspired epistemology to metaphysics signals a first displacement of the transcendental imagination. The role of the subject is shifted from standing as the very source of the faculty of the imagination to a mere moment or conduit in the process of imaging. In view of the centrality of the imagination in Fichte's work, and his growing interest in magnetic psychology's discourse of subtle matter and its power synthesis, one is led to wonder whether Fichte's tentative embrace of "subtle matter" does not—as a possible second displacement of the imagination—signal something like a material imagination. After all, magnetic psychology's willful manipulation of "subtle matter" was a material technique that could direct an imagination that now seemed less a faculty than something quasi-material.

The present chapter establishes the background for such greater claims. In this chapter, I will show how Fichte's transformation of his understanding of transcendental philosophy to require pragmatic verification led him to seek an empirical proof for his transcendental claims.[7] These proofs led him most often to that quasi-material phenomenon which best materially instantiates the Wissenschaftslehre's striving for a unification of ideal and real worlds—subtle matter.[8] In brief, Fichte's transcendental account seeks proof in the phenomenon of magnetic rapport and subtle matter. Here, however, we will explore precisely how the transcendental is transformed, and how and why a historico-empirical verification becomes necessary. Additionally, I attempt to develop how the idea of subtle matter and its account of influence begins to transform transcendental idealism's account of the social.

The significance of this chapter in its own right is to try to understand how this reinterpretation of Fichte's work, in light of his references to subtle matter, alters our very understanding of his account of the social. In the latter part of this chapter, by expanding our interpretation of the sort of influence subtle matter represents in Fichte's work—through the idea of aesthetic influence in Kant's *Critique of Judgment*—I try to analyze just what sort of account Fichte's description of subtle matter's role in intersubjectivity is intended to represent. In other words, I will try to develop Fichte's rather

6. See Dreschler, *Fichtes Lehre vom Bild*, and Janke, *Fichte: Sein und Reflexion*.
7. See Dreschler, *Fichtes Lehre vom Bild*, and Janke, *Fichte: Sein und Reflexion*.
8. Fichte seeks such verification in subtle matter, I will show, in both the *Grundlage des Naturrechts* and the "Tagebüch über den animalischen Magnetismus."

obscure account of the material/empirical dynamic of *Aufforderung* and *Aner-kennung* in and through the terminology "intellectual influence" and "subtle matter." And these latter terms, in turn, will be brought to relief with the help of the interpretative horizon of Kant's aesthetics.

The Fichtean Transcendental Method

For Kant, transcendental philosophy offered us an a priori mode of knowl-edge, a "knowledge which is occupied not so much with objects as with the mode of our knowledge of objects insofar as this mode of knowledge of objects is possible a priori" (*CPR* B25). Yet while Fichte continued to carry out the spirit of Kantian philosophy, he transformed the transcendental method in two important respects.[9] First, Fichte was less concerned with the knowledge of an object, understood as a product constituted in and through the transcendental method, than with the activity or process of the transcen-dental method itself. For Fichte, one does not take part in the transcendental project by becoming a disciple of Kant or merely reading him. Rather, its method is a continually developing project, a project or activity that one must engage in on one's own. In fact, in the "Second Introduction to the Wissenschaftslehre" Fichte explicitly refers to the transcendental investiga-tion in terms of an "experiment" (*IWL* 37). He explains that while other philosophies, like Kantian philosophy, "consist entirely in . . . arguments that they themselves construct," and thus, in short, traffic in "dead concepts," "the Wissenschaftslehre is something vital and active, something that gener-ates cognitions out of itself and by means of itself" (*IWL* 36–37). In fact, the "First Introduction to the Wissenschaftslehre" begins, quite tellingly, with an imperative to take up philosophy and transcendental philosophy as an *experimental practice*. Fichte implores: "Attend to yourself; turn your gaze from everything surrounding you and look within yourself: this is the first demand philosophy makes to anyone who studies it" (*IWL* 7).

Fichte's emphasis on transcendental philosophy as a first-person, experi-mental method also transformed the scope and limits of Kantian-style episte-mology. For instance, whereas Kant showed us that our own cognitive structure plays a central role in the ontological constitution of objects, Fichte

---

9. Fichte, of course, saw himself as the only true Kantian. He writes that the Wissenschaftslehre "is nothing other than the Kantian philosophy properly understood" (*IWL* 52).

could not adhere to such a rigorous interpretation. Thus, although concepts without intuitions are blind, for Fichte, the legislative power of the Kantian transcendental a priori is largely a construction of thought in which validity is either affirmed or denied at the level of the transcendental deduction. Fichte's emphasis on the experimental activity involved in the transcendental method, however, describes the transcendental model as a mere hypothesis that depends upon a posttranscendental move. In other words, because the Fichtean ego itself has no ontological status, and is not even a substratum, it stands as a mere hypothesis (*FW* 1:515). The logical principles of identity, difference, and sufficient reason are given coherence through the postulation of a hypothetical ego whose unifying activities of thesis, antithesis, and synthesis amount to a speculative unity, ought, or ideal, which ultimately cannot operate without reference to the real.[10] This mutual dependency of the theoretical and practical series within Fichte's work offers a new understanding of the role of transcendental constitution in which transcendental confirmation is no longer wholly independent of empirical considerations. While the epistemological claims of Kantian transcendental philosophy could be sorted out at the transcendental level, the hypothetical structure of the Fichtean transcendental requires a "post-transcendental confirmation."[11] In order to understand the nature of this confirmation, however, we must first understand Fichte's conception of the relation between the empirical and the transcendental.

Admittedly, the transcendental perspective would seem to have little to do with empirical concerns. After all, in the "First Introduction to the Wissenschaftslehre" Fichte rhetorically asks, "How could anything that limited itself to experience be called philosophy?" Fichte suggests philosophy, and in particular transcendental philosophy, is something wholly independent of experience. Transcendental philosophy is something which determines and gives shape to experience, not the reverse. Thus, Fichte warns, "Those who advise you to keep your eye constantly focused upon experience when you philosophize and advising you to fudge its factors . . . as procedure {is} as dishonest as it is superficial" (*IWL* 33). In other words, one should not allow experience to determine transcendental philosophy: such a procedure is a sham. Yet what is so perplexing about this position Fichte establishes is that while empirical experience cannot and should not determine the content of transcendental

<hr />

10. Williams, *Recognition,* 40.
11. Ibid.

claims, it nevertheless can determine their validity. Fichte writes: "If the results of such philosophy do not agree with experience, then the philosophy given is surely false" (*IWL* 32). In other words, *it is with reference to experience that transcendental claims are falsifiable.* While, of course, Fichte is of the opinion that experience would almost always conform to and thus stand as the ultimate proof of a procedurally rigorous transcendental philosophy—like that carried out in the Wissenschaftslehre—what happens when transcendental philosophy does indeed need correction through reference to experience? Is this not an instance in which transcendental philosophy keeps its eye on experience? Further, if this new method of transcendental philosophy requires a posttranscendental verification, then it would seem to be a procedure that to some degree is determined precisely by empirical concerns. The nature of the relation of the empirical and transcendental in Fichte's philosophy needs to be further clarified through an analysis of these two orders as they arise in Fichte's concept of the Wissenschaftslehre. These two orders, the empirical and the transcendental, are approached by Fichte through his exploration of the dual series of real and the ideal, and his understanding of the relation between intellectual and bodily concerns. We begin with the real and the ideal.

The Dual Series: The Real and the Ideal

Fichte explains in the "Second Introduction to the Wissenschaftslehre" that "the Wissenschaftslehre contains two very different series of mental acting [*des geistigen Handelns*]: that of the I the philosopher is observing, as well as the series consisting of the philosopher's own observations" (*IWL* 37). What Fichte is distinguishing are two different orders of experience, ordinary consciousness and philosophical consciousness. And the failure to recognize the mutual existence and dependence of these two orders, Fichte warns, "is one of the main reasons the Wissenschaftslehre has been misunderstood" (*IWL* 37–38). What Fichte is suggesting is that while on the one hand ordinary consciousness cannot be readily understood without the clarifying theoretical apparatus of transcendental philosophy, so too would transcendental philosophy be a pure logical abstraction without the pragmatic verification in practical experience.

There have been two central interpretations of the relation between these orders. The first, that represented for instance by the work of Günter Zöller and others, argues that for Fichte the real is essentially still an intellectual category. Zöller writes, "The sense of 'practice' (Praxis) introduced in Fichte's account of the I as originally practical . . . is decidedly intellectual."[12] While there is much evidence for this view, the danger of embracing its particular interpretative tack is that, in emphasizing the intellectual form of the real, it tends to collapse the two orders, reducing the real to the ideal, and thereby obscuring the unique role of the empirical in Fichte's new account of the transcendental. The other view, that represented by commentators like Robert R. Williams, argues that the real and the ideal must truly stand as two orders if the Fichtean transcendental is not to "be a closed system of thought determinations measured simply by the criterion of coherence."[13] In other words, although the ideal stands as an "ought" in Fichte's system, it is a task that is not yet accomplished. And the correlation between the transcendental and empirical is such that while the ideal has established the task, the transcendental order is nevertheless "subject to modification, if not correction, by the historical order."[14] While there is much evidence for both views, Williams further—and in my view rightly—emphasizes the impossibility of such a choice. He writes, "These two standpoints can neither be reduced to the other nor reduced to a superior unity. They cannot be finally unified or identified. Since the ideal and real ground cannot be identified, the transcendental that emerges in the Wissenschaftslehre is one that requires history."[15]

Ordinary consciousness may require philosophical consciousness for its explanation, but the transcendental method of philosophical consciousness (in turn) remains existentially rooted within the concerns of practical experience. After all, for Fichte, one recalls, first principles cannot be proven. The choice between idealism and dogmatism was motivated by practical concerns—specifically, the concern for freedom (*IWL* 15–18). And it is in this sense that the ideal and real stand as a dual series whose seeming independence is actually constituted in an implicit mutual dependence. Now Fichte's complaint that dogmatism is a realism, a materialism, and in short, a fatalism (*IWL* 16) stems from his belief that dogmatism denounces human freedom in its very principles by reducing the free will of the human intellect to the principle of causality in which it becomes a mere pawn in the structure of a larger causal series (*IWL* 22). He is lamenting dogmatism's disregard for the

12. Zöller, *Fichte's Transcendental Philosophy*, 79.
13. Williams, *Recognition*, 41.
14. Ibid.
15. Ibid.

practical choice for freedom that ought to ground otherwise undecidable first principles (*IWL* 22). With dogmatism, the "real" series of causality is one that reduces freedom and spirit to a mere causal explanation. In a more contemporary register, for example, such an approach would reduce the phenomena of mind to the materialist discourse of the physical anatomy of the brain. Yet, like the preservative force of Cartesian dualism against the onslaught of materialist explanation, Fichte's dual series of the real and the ideal rejects the purely causal explanation on practical grounds—for the sake of freedom.

In the first part of the *Wissenschaftslehre* of 1794 Fichte presents a genetic history of consciousness that reconstructs the developmental history of consciousness through its logical categories. As I suggested, to the extent that this transcendental reconstruction is not directly constitutive of experience (as it was for Kant), it stands as a mere hypothesis that awaits a practical negative verification in the form of falsification. Here we see the Wissenschaftslehre's double series at work. Philosophical consciousness must stand in conjunction with ordinary consciousness; the theoretical must reference the practical, and vice-versa. If, in anticipation of Hegel's *Phenomenology of Spirit,* Fichte's transcendental method recognizes the developmental and historical character of the transcendental categories, the real question is the following: what is the relation between this hypothetical transcendental analysis of consciousness and the actual ordering of pragmatic history?[16]

Two options stand before us. The first, a prospective relation, would understand the transcendental as a hypothesis that awaits a practical or empirico-historical verification. The second, a retrospective relation, would hold that transcendental reflection, like all philosophical experience, is merely a second-order reflection that can only come after or follow experience.[17] In fact, both versions seem to exist within the Wissenschaftslehre. The first understands the tension between the theoretical and empirical (or ideal and real) series from the perspective of the theoretical, while the second views this tension from the perspective of the practical. Still, Fichte's explanation of this dual series is not entirely satisfactory. While in articles like the review of *Aenesidemus* Fichte explains that the theoretical and practical aspects of this dual series stand in a circular relation (*EPW* 75–76), so that, for instance, with the termination of the ascent of the theoretical, begins the descent into

16. Ibid.
17. Ibid., 40.

the practical, and so on, at other times Fichte suggests that no logical ordering or coincidence of this series is even possible. Why this is so requires us to return to address more explicitly the status of the transcendental in Fichte's work.

This tension between the practical and the theoretical (that forms the very essence of the Wissenschaftslehre) is also the very source of Fichte's seemingly unusual explanation of the transcendental and its relation to the empirical. Transcendental philosophy ought to be a closed system and complete unto itself. However, to the extent that transcendental philosophy is rooted in ordinary consciousness and, according to Fichte, cannot stand as an abstract system—insofar as the real cannot be reduced to the ideal—it requires a posttranscendental falsification. The empirical world, it would seem, can falsify or disprove transcendental claims.

In the Wissenschaftslehre the practical seems to guide the theoretical from the outset. Insisting that first principles cannot be derived logically, but by practical demands alone, Fichte recognizes that the interests of ordinary consciousness guide and establish the horizon and concern of the theoretical. Philosophy itself then begins with a pragmatic decision. Yet although the Wissenschaftslehre begins by recognizing the general undecidability of first principles, and the consequent priority of the practical, the unity of this double series—that is, the unity of the real and ideal—itself nevertheless stands as a practical demand—as an "ought." For Fichte, this striving for unity is the key to understanding how for Kant reason can be practical. In the review of *Aenesidemus* he explains: "In the pure ego reason . . . is practical only insofar as it strives to unite the two . . . a practical philosophy emerges by going through in descending order the stages which one must ascend in theoretical philosophy."[18] Fichte began this ascent in the first half of the *Wissenschaftslehre* of 1794 through a genetic reconstruction of the logical categories of consciousness. This reconstruction, however, was merely a hypothesis that would demand falsification within the practical part. In other words, the descent into the practical retraces the theoretical ascent in reverse order as a practical or historico-empirical falsification of the transcendental hypotheses. The precise relation between ordinary and philosophical consciousness, between the truth of pragmatic history and that of transcendental philosophy, is not easy to sort out. There is no one to one correspondence between the elements of the real series and those of the ideal series. As a

18. Ibid.

consequence, what is established is a tentative correlation that prevents either account from affirming a closure, or truth. From the context of the spirit of Kantian philosophy, Fichte seems to embrace truth in the form of transcendental truth. Nevertheless, one of the practical demands of an ordinary, historical, empirical consciousness is the call for a revision of this transcendental truth, so that the Fichtean transcendental cannot be conceived apart from the practical demands of a pragmatic history of spirit.[19] Thus, as we will see both in Chapter 6 with the analysis of subtle matter in the *Grundlage des Naturrechts,* and in Chapter 8 through an exploration of the "Tagebuch über den animalischen Magentismus," while Fichte thinks that history and the empirical world will largely bear out the hypothesis established by the transcendental method, it often does not. Fichte must begrudgingly recognize and embrace just what his very system of the Wissenschaftslehre offers. In other words, although the empirical world does not directly dictate the level of transcendental analysis, pragmatic history, and the practical historico-empircal demands upon the finite subject, it does necessarily loosen the scope, structure, and claims of transcendental philosophy as traditionally understood in the Kantian sense.[20] The world is no longer understood as the direct effect of transcendental constitution: transcendental claims now require a pragmatic falsification as a litmus test of their validity. Still, neither the real nor the ideal, neither the empirical nor the transcendental, can be reduced to the other. Their unity remains an "ought": it is a practical demand on the order of the real that stands simultaneously as the culminating point of the ideal.

To the extent that to embrace either side apart from the other is mistake, Fichte will conclude in the *Wissenschaftslehre* of 1794 that taken individually "both views are wrong." As a consequence, one can embrace neither transcendental philosophy in the Kantian sense, nor dogmatism in the form of materialism, empiricism, or realism. With the insight that "the second series is no new series at all," but "merely the first in reverse" (*FW* 1:224; *SK* 200), Fichte asks the penetrating question of "how ideality and reality can be one and the same; how they differ only in the manner regarding them, and how the one can be inferred from the other" (*FW* 1:226; *SK* 202). Fichte realizes that what he is left with is a gap between "incapacity and demand." He explains that "the mind lingers in this conflict and wavers between the two—

19. Ibid., 37–43.
20. Ibid., 41.

wavers between the requirement and the impossibility of carrying it out" (*FW* 1:225; *SK* 201).

Yet it is out of this gap—the gap between the practical *demand* for unity and the *incapacity* of transcendental, theoretical consciousness to wholly account for its own conditions of possibility in ordinary consciousness and the lifeworld—that a new productive power arises: it is out of the tension between incapacity and demand that what Fichte calls the productive power of the imagination first appears (*FW* 1:226; *SK* 201). A central question for this project is whether this productive power of the imagination is itself exclusively a transcendental faculty, or whether, as a power that emerged out of the tension between the real and ideal, it stands somehow beyond such characterizations, since it is itself a power of synthesis. In other words, if the task of the Wissenschaftslehre is somehow to move beyond the dual series of the real and the ideal by means of the synthetic power of the imagination, then it seems the achievement of such a quest would also require that the imagination itself stand beyond this dual series. It would entail that Fichte articulate the productive power of synthesis, called the imagination, beyond an exclusively transcendental framework. What is called for is a new concept of the imagination.

In Chapters 7 and 8 our analysis of the imagination in the *Wissenschaftslehre* of 1813 will bear out such out suggestions. There I show that Fichte recognizes that the tension between the demand for unity and its theoretical incapacity—achieved momentarily through the synthetic power of the imagination—can itself no longer be conceived as a strictly theoretical or transcendental faculty. Here, the dual demand in the conflict between the real and the ideal requires that the imagination itself function as a third term. Fichte is in the difficult—if not impossible—position of attempting to unite the ideal and real series through the terms of imagination and subtle matter, all while recognizing that his account of such a unity must nevertheless take place from the perspective of the real or the ideal. Although Fichte uses the terms "imagination" and "subtle matter" as a means to move away from this dual series, one must also recognize that it is the imagination and subtle matter that form the ideal and real aspects (respectively), and thus together stand as related binary aspects of yet another dual series.[21] Nevertheless, like the very dual series it hoped to move beyond, the relation of the imagination

---

21. What I am suggesting is that something like a nested "dual series" exists within Fichte's work.

to subtle matter reinscribes this series so that subtle matter will stand as a material, empirical proof for the imagination's claims of synthesis.

And it is in this sense that Fichte's own quest for a resolution between idealism and dogmatism (the real and the ideal) in and through the imagination leads him to the notion of "subtle matter." Fichte comes to recognize that his own philosophical quest for a power of synthesis in the imagination is recapitulated historically both in the conflict between idealism and materialism and in quest for a third term in the notion of subtle matter. Fichte, however, will first invoke the notion of subtle matter as a third term in the *Grundlage des Naturrechts* in an attempt to explain the relation between spiritual and material influence. And it is to the *Grundlage des Naturrechts* to which we must now turn.

The Transcendental and the Empirical in the *Grundlage des Naturrechts*

While Fichte is clearly engaged in transcendental philosophy in his deduction of the transcendental conditions of intersubjctively constituted right and freedom in the *Grundlage des Naturrechts,* our previous analysis of Fichte's development of the transcendental as something empirically falsifiable should help give us insight into why Fichte would ever need to invoke something like subtle matter at all.

A continued exploration of the role of the dual series in Fichte's thought should give us a clear indication why Fichte becomes concerned in the *Grundlage des Naturrechts* not merely with the transcendental deduction of the other, but with the actual, concrete other, and the need to explain that concrete reciprocal influence by reference to subtle matter. Now the complex relation of this dual series to the other was first made visible in the *Wissenschaftslehre* of 1794 with Fichte's articulation of the *Anstoß* or Not-I.[22] As a consequence of this dual series, two different accounts of or perspectives on the *Anstoß* are possible. From the theoretical or transcendental perspective, the other is interpreted idealistically as a mere self-restriction of the absolute I. However, from the practical perspective of ordinary consciousness, the other is an actual other upon which I depend for freedom and self-conscious-

---

22. These terms actually reference different moments. The Not-I is the conceptual articulation of the *Anstoß.* I detail this later in this chapter.

ness. The greater question that needs to be asked now is how this dual series comes to be established within the *Grundlage des Naturrechts*. One might also ask this same question by asking how the very notion of the *Anstoß* comes to be reformulated in the *Grundlage des Naturrechts*.

Just as this dual series appeared in the Wissenschaftslehre in its discourse of the *Anstoß*, so too does this dual series reappear in the *Grundlage des Naturrechts* through the discourse of the *Aufforderung*. In what follows I make this argument by outlining Fichte's basic aspirations of the transcendental deduction of the other in the *Grundlage des Naturrechts*, and then show how he attempts to explain a concrete or actual influence among beings through reference to the discourse of subtle matter. I detail the relation between the sort of influence hypothesized in the transcendental deduction of right and the postulation of a material, affective influence, through a detailed analysis of the vocabulary of *Anstoß, Aufforderung*, and *Anerkennung*. Before I turn to such an analysis, however, I want to give a very brief overview of the task of the *Grundlage des Naturrechts*.

Reassessing the Task of the *Grundlage des Naturrechts*

The central task of the *Grundlage des Naturrechts* is to explain how freedom is possible. Freedom, for Fichte, is possible through a mutual self-limitation, a self-limitation achieved through the mutual recognition of an equal right to freedom. Thus it is a recognition of the rights of the other that makes freedom possible. Fichte explains that "right" is constituted intersubjectively. Thus, the relation of right (*Rechtsverhältnis*) is grounded in the fact that "each limits his own freedom through the possibility of the freedom of the other" (*FW* 3:120–21). Fichte arrives at such insights through a transcendental deduction of right. In other words, a recognition of the other and his or her rights stands as a transcendental condition of possibility of an autonomous subject—which defines the self-active ego. As paradoxical as this may sound, Fichte is explaining how autonomy is indeed possible only by virtue of the self-conscious subject's dependence upon the other. But in what respect is the subject dependent upon this "other"?

The other stands as a catalyst to an awareness, a self-conscious awareness of the subject's power of self-determination, as a freely determining self with a determinate sphere of influence. The other catalyzes the self's awareness of his or her own capacity for free action and freedom. In the deduction of right, Fichte explains that it is through the inference that this other, like me, is a free self-determining being—that I recognize his or her right to free activity and as such limit my own action. This freely determined self-limitation constitutes the freely limited sphere of activity known as the subject. It is through a recognition of the other, through this intersubjective recognition, that I limit my own activity and define myself as a self-delimited, freely determined, and thus self-determining being. Summarized then, the sphere of right delimits that sphere of action called the ego. That the solidity of one's sense of self is tied to a sphere of rights is perhaps witnessed no more clearly than the loss of self in the destruction of one's essential "human rights" through the phenomena of torture and solitary confinement. Here, without intersubjective recognition, one's very sense of self-consciousness is threatened.

One central problem that remains is that while Fichte's method of deducing the other as transcendentally necessary seems quite sound, he nevertheless has much more difficulty explaining *how* our relation to the other is *actually* constituted. And while one might argue that Fichte's project is exclusively transcendental and consequently he is not concerned (or ought not to be concerned) with the empirical details of how intersubjective influence actually takes place, I would offer two counterarguments. First, Fichte's reformulation of the Kantian transcendental—as one which requires a post-transcendental move—suggests that Fichte needs to be concerned with the empirical falsifiability of his transcendental claims. And second, that Fichte's description of subtle matter in the *Grundlage des Naturrechts* is critical proof that he was indeed concerned with delimiting the concrete details of what—at first glance—appears to be a exclusively transcendental argument. I will begin exploring the precise relation of the transcendental and the empirical in the *Grundlage des Naturrechts* through a detailed development of its terminology.

Aufforderung and Anerkennung

If the other is not a real concrete other, but merely a hypothetical other, deduced as a transcendental condition of my own ascension to freedom and self-consciousness, how precisely does Fichte understand the terminology of recognition (*Anerkennung*) and the summons (*Aufforderung*) to operate? While it will be important to understand the relation of the *Anstoß* to the

Aufforderung (and thus the relation of the Wissenschaftslehre to the *Grundlage des Naturrechts*) in order to see whether the dual series reappears yet again in the *Grundlage des Naturrechts*'s account of the other, for the moment I would like to begin with an initial description of the *Aufforderung* and *Anerkennung*. Now although the *Aufforderung* and *Anerkennung* are often understood to be terms that essentially support a transcendental deduction of a hypothetical other, I argue, through reference to several significant scholars, that the *Aufforderung* operates in a weaker transcendental sense than is generally recognized, and that the *Anerkennung*, in fact, is subject to a concrete, reciprocal, quasi-empirical falsifiability. In short, I will argue that Fichte's account of right in the *Grundlage des Naturrechts* requires a posttranscendental move. And while his very terminology moves him toward such an end, his references to the dynamic of subtle matter as an attempt to explain the empirical details of the influence the *Aufforderung* offers the ultimate proof of his desire to seek a verification of his transcendental claims.

In his description of the *Aufforderung* in section 3 of the *Grundlage des Naturrechts,* Fichte seems to engage in a regressive transcendental-style argument. The lead assertion is as follows: "A finite rational being cannot ascribe to itself free efficacious action [*freie Wirksamkeit*] in the sense world, without also ascribing freedom to others, and therefore without also assuming [*anzunehmen*] other rational beings besides itself" (*FW* 3:30). Fichte is telling us that if one is to ascribe freedom to oneself as a rational being, this is only possible, deduced as transcendentally necessary, insofar as such characteristics are also ascribed to others. Yet is Fichte merely engaged in a transcendental argument that points to the other as transcendentally necessary for my own freedom and self-consciousness?

Edith Düsing, C. K. Hunter, and Robert R. Williams, among others, have all suggested that the *Aufforderung* is *not* something deduced as transcendentally necessary; but rather, it itself exists as a transcendental fact.[23] As Williams argues, "It is not something inferred as ground, but a fact or starting point. As such, it refers to the prior action of the other. Fichte's discussion is not so much a transcendental argument as it is a description and analysis of the fact of the *Aufforderung,* i.e. the fact of being summoned."[24] The summons or demand (*Aufforderung*) of the other is a fact—albeit a contingent

---

23. Düsing describes it as a "necessary fact." See Düsing, *Intersubjektivität und Selbstbewußtsein,* 247; Hunter, *Der Interpersonalitätsbeweis;* and Williams, *Recognition,* 60.
24. Williams, *Recognition,* 60.

fact—upon which our very freedom and self-consciousness depends. Here, to the extent that the *Aufforderung* is not simply the product of a deduction, but is itself recognized as a transcendental fact, suggests that there must be an actual—and not merely hypothetical other—upon which my freedom and self-consciousness depends. The paradox, of course, is that Fichte largely arrives at such conclusions by means of a transcendental argument.

Now while the requirement of the other as a transcendental fact stands as a formal condition, the notion of *Anerkennung* is more concrete.[25] The *Aufforderung* describes an asymmetrical relation whereby I remain dependent upon this contingent fact for my own achievement of freedom and self-consciousness. *Anerkennung,* by contrast, is a concrete reciprocal relation whereby I am both a receiver and a conferrer of recognition. Fichte describes the concrete relation among beings in the dynamic of recognition as follows: "Neither can recognize the other if both do not mutually recognize each other. And neither can trust the other as a free being if both do not do so mutually and reciprocally" (*FW* 3:41).[26] The point we should gain from this is that while with the *Aufforderung* we remain dependent upon what Düsing calls the "necessary fact" of the other, with the *Anerkennung* we stand in a real reciprocal relation to the other. Proof that we are not speaking of a mere concept is clear in Fichte's emphasis upon "action." Fichte explains: "The entire described union of concepts was only possible in and through action [*Handlung*]. . . . Actions instead of concepts are what really matter here. And there is no talk here of mere concepts without corresponding actions, for, strictly speaking there could be no talk about such at all" (*FW* 3:45).[27] Fichte's point is that it is not enough that I have been able to account for my own epistemological status as a self-conscious being capable of free action— through a transcendental deduction of the other as a mere construct or concept. Rather, insofar as the *Grundlage des Naturrechts* is as much a work concerned with ethical action—from a man who defined the ego as act—as it is concerned with law and right.

From *Anstoß* to *Aufforderung:* A Dual Series in the *Grundlage des Naturrechts?*

The issue of sorting out the transcendental and the empirical is no less easy here in the *Grundlage des Naturrechts* than it was in the Wissenschaftslehre.

25. Ibid., 61.
26. Cited in Williams, *Recognition,* 61.
27. Cited in ibid., 62.

With this brief introduction to the terminology of *Aufforderung* and *Anerkennung* (and the suggestion that the *Anerkennung* involves a truly concrete reciprocity, while the *Aufforderung* implies a modified and weaker sense of the transcendental), I would like to begin to assess how or to what extent the Wissenschaftslehre's dual series can be understood to be inscribed in the methodology of the *Grundlage des Naturrechts*. The confirmation of such a claim would show that, like the Wissenschaftslehre, the *Grundlage des Naturrechts* does not rest exclusively on a transcendental argument; but rather, seeks the post-epistemic move of falsifiability, both through its own terminology of *Anerkennung*, and through reference to the discourse of subtle matter. In order to develop such a claim, we must begin to understand more fully the relation between the *Anstoß* and the *Aufforderung*. To do so I will begin to outline the relation between the Wissenschaftslehre and the *Grundlage des Naturrechts* through an overview of some of the scholarly literature on the relation between the Wissenschaftslehre's *Anstoß* and the *Grundlage des Naturrechts*'s *Aufforderung*.

Although the *Grundlage des Naturrechts*'s notion of the *Aufforderung* has much in common with the Wissenschaftslehre's understanding of the *Anstoß*, their precise relation is not easy to grasp. Fichte understands the *Anstoß* to be a "first mover" (*erste Bewegung*) (*FW* 1:279). For instance, when Fichte claims in the "First Introduction to the Wissenschaftslehre" that the ego is self-determining, but not self-creating, what he is suggesting is that some external other or catalyst of some sort is necessary (*FW* 1:279). It is not the ego and the ego alone that is the source of everything. While we might invoke the notion of the dual series again to show to that this very *Anstoß* can be read idealistically (as a self-imposed limitation), or realistically (as an external shock), either perspective makes clear that for Fichte the *Anstoß* functions as at least some form of limitation of the ego.

Following out the logical consequences of Edith Düsing's assertion that the *Aufforderung* is a transcendental fact, one wonders whether or not the *Anstoß* is also a transcendental fact. Fichte answers: "The *Anstoß* too, does not stand as a mere condition of possibility; but rather, as a transcendental fact, insofar as it would never have been catalyzed to action or even existed without such a first mover" (*FW* 1:279).[28] Yet if the *Anstoß* and *Aufforderung* can both be understood as "transcendental facts," what is their relation? Does one term merely represent the continued development of the other?

Edith Düsing sees a significant distinction between these two terms. For

28. Ibid., 57.

her, the distinction between the *Anstoß* and the *Aufforderung* marks a division between the Wissenschaftslehre's transcendental account of the pure ego and the *Grundlage des Naturrechts*'s description of the empirical ego. She writes: "In clear distinction to the transcendental mode of explanation, the *Anstoß* in the *Grundlage des Naturrechts* signifies an empirical event in time and space. Through this *Anstoß* the concrete subject finds itself determined by a reality outside itself."[29] Yet while Düsing claims there is a sharp distinction between the empirical and the transcendental ego, textual support for such a claim seems less than clear. Still, others have argued a different means for establishing transcendental "levels" in Fichte's work.[30] For instance, like Düsing, Ludwig Siep has argued that the Wissenschaftslehre stands as a foundation for the *Grundlage des Naturrechts,* which functions something like an application or extension of the former.[31] Wilhelm Weischedel refines Siep's account through the language of ontology by suggesting that the *Grundlage des Naturrechts* functions as a regional ontology of the Wissenschaftslehre's more general ontology.[32] Such an analysis suggests that Fichte's project of the Wissenschaftslehre and the *Grundlage des Naturrechts* stand in reciprocal relation, much like the dual series. One stands as a practical application of what was formerly only a theoretical or transcendental account. Yet insofar as both the *Grundlage des Naturrechts* and the Wissenschaftslehre seem to have (additionally) a dual series within them, are we then dealing with an extremely complex nested cluster of dual series?

In direct contrast to these aforementioned accounts, Peter Baumann argues that the *Grundlage des Naturrechts* cannot stand as an instance or application of the Wissenschaftslehre because the Wissenschaftslehre offers no deduction of the social.[33] Still other scholars like Reinhardt Lauth and Alexis Philonenko argue that the notion of the social was already present at the time of Fichte's first conception of the Wissenschaftslehre.[34] With reference to Fichte's *Die Bestimmung des Gelehrten,* Lauth suggests that for Fichte the issue of intersubjectivity was not an afterthought but, in fact, the central question of philosophy that Fichte would develop through future revisions of the Wissenschaftslehre.[35] Since, as scholars like Reinhardt Lauth and Wolf-

29. Düsing, *Intersubjektivität und Selbstbewußtsein.*
30. The literature review that follows is in part indebted to Robert William's own review of the literature.
31. See Siep, *Anerkennung.*
32. Weischedel, *Der frühe Fichte.*
33. See Baumann, *Fichtes Ursprungliches System.*
34. See Philonenko's *La liberté humaine dans la philosophie de Fichte* (Paris: PUF, 1966), and Lauth's "Le problème de l'interpersonalité chez J. G. Fichte."
35. Ibid.

gang Janke have shown that Fichte's work ultimately develops in such a way that the transcendental subject gradually loses its certainty before the other and being, so that it comes to function as a pure mediation (*Durch*) of the social, one might understand the *Aufforderung* as merely a tentative development of the *Anstoß* in Fichte's attempt to understand the place or situatedness of the transcendental subject in its social world.

This, of course, is not to say there is no significant difference between these terms. The most significant difference for this project is that while the *Anstoß* functions in a seemingly causal manner in which it is something felt, but not known, the *Aufforderung*, by contrast, demands the component reciprocity of recognition (*Anerkennung*)—a sort of knowing in acknowledgment. In what follows, I would like to develop these terms from the point of view of their respective forms of influence. In other words, if the *Anstoß* seems to function on the order of the causal model, how might we understand the dynamic of *Aufforderung*? After all, a recognition of right is certainly not a causal phenomenon, but something that arises through recognition, or in Fichte's words, through an intellectual influence. Yet, just what is the nature of intellectual influence?

From our discussion of both the idea of the *Aufforderung* as a transcendental fact and Fichte's discussion of the dual series, and the need for a post-epistemic empirical falsifiability of transcendental claims, it should be clear that our discussion of "influence" (*Einwirkung*) need not be limited to a transcendental hypothesis. In fact, as I will argue, Fichte will ultimately attempt to develop in detail the actual dynamic of the influence involved in the *Aufforderung*—as a post-epistemic proof—through his discussion of subtle matter.[36] As a lead into this unique form of influence, we will begin with a discussion of the *Anstoß* and the causal model of influence.

The Question of Influence: From *Anstoß*, to *Aufforderung*, to Subtle Matter

Just what kind of influence (*Einwirkung*) is represented by the terminology of *Anstoß* and *Aufforderung*? One might argue that since these terms are tran-

36. While I have already outlined in an earlier footnote that Fichte's demand for a posttranscendental move to support his transcendental claims tended to waffle between, on the one hand, his aspiration for *Wissenschaft* in the form of an attempted *falsification* of transcendental claims and, on the other, his more fantasy-like search for an empirical proof or *verification* of transcendental claims, I believe that his references to subtle matter clearly fall into the latter category.

scendental facts and necessary conditions for experience, no detailed explanation of how influence actually occurs is needed; I contend that such a conclusion ignores Fichte's reformulated notion of the transcendental and its demand for a post-epistemic proof. Specifically, I will show that Fichte's own attempt to explain the details of the dynamic of *Aufforderung* and *Anerkennung* through the terminology of subtle matter attests not only to the need for a posttranscendental proof, but additionally to Fichte's desire for a theoretical understanding of how the actual dynamic of influence occurs.

Now although the *Anstoß* and *Aufforderung* are often conceived as transcendental facts about which little more can be said, the received reading of Fichte understands the influence of the *Anstoß* to operate on a causal model. From this interpretative tack, little is said about the type of influence represented by the *Aufforderung*. But what about the *Aufforderung*? Surely the demand of the other does not exert a causal-like influence upon me? What then is the nature of this influence? Insofar as I understand the *Aufforderung* as an extension and development of the notion of the *Anstoß*, I will begin to explore the kind of influence involved in the *Aufforderung* by expanding the canonical reading of the *Anstoß* as a mere causal phenomenon to show that the *Anstoß*, and by extension the *Aufforderung*, is perhaps best explained not in terms of causality, but from the perspective of affection or feeling. After all, this is why in the previous chapters we spent so much effort establishing the significance of self-affection as a possible resolution to the so-called paradox of self-reflection.

Fichte's attempt to explain how the actual dynamic of intersubjective recognition of right occurs forces him (once again) to confront the divide between material and immaterial worlds. After all, how could a causal, material influence produce an intellectual or spiritual influence—and vice-versa? This is the real crisis of influence that Fichte is forced to confront. If the *Anstoß* and *Aufforderung* are to be of any real value, they must somehow operate not merely within a causal paradigm; they must additionally bridge the divide between material and intellectual worlds. I will suggest that Fichte begins to expand the limits of this terminology beyond the *Anstoß*'s initial causal framework through an emphasis on the notion of feeling and affection, and through his eventual turn to the discourse of subtle matter.

The *Anstoß*: From Causality to Affect

The English translations "check" and "limitation" do not do justice to the complexity of the term *Anstoß*. In response to the term's obscurity Pierre-

Philippe Druet has suggested that Fichte's term *Anstoß* stands as a metaphysical rendering of the problem of bodily interaction as posed by the school of rational mechanics in the post-Cartesian tradition.[37] Such an interpretation seems to bear itself out with Fichte's remarks from *The System of Ethics* (*Die Sittenlehre*) (1798). In this work he writes: "If, for example, a ball is put into movement by a check [*einen Stoss*], I can thus observe immediately the movement of the ball, the point from which it proceeds [*ausgeht*], and thus the speed with which it is moved. . . . But I am able to deduce [*schliessen*] all of this from the force with which the ball has been hit [*mit welchem sie angestossen ist*]" (*FW* 4:24).[38] Fichte's reference to billiards makes clear that the notion of the *Anstoß* had its origins in the paradigm of causal, physical impact. Nevertheless, what is really at issue is the influence of one body upon another. The *Anstoß* is one such explanation. While ultimately for Fichte—and much of nineteenth-century science—the real paradox will turn out to be the problem of action at a distance, as witnessed, for example, with objects, through the phenomenon of magnetism, and by analogy, with people, through the phenomenon of magnetic rapport. Before we take up the problem of action at a distance, however, we need to return to the problem of the mere interaction of bodies as described by the *Anstoß*. It is with the *Aufforderung* and Fichte's appeal to subtle matter in his explanation of its communicative dynamic that the issue of action at a distance is confronted.

At times Fichte seems to associate the *Anstoß* with that limitation of the I referred to as the Not-I. As we saw in the dynamic between the I and the Not-I, the original activity of striving was limited and forced back upon itself. Here, however, with talk of self-limitation in which the I posits the Not-I—in opposition to the I—we are already witnessing an intellectual construct. An already delimited subject takes hold of its capacity for self-determination by accounting for its own limitation as "self-determined." The *Anstoß*, however, is the prior moment. Like the noumenal subject, the *Anstoß* is *not* to be associated with a concept, with a concept of the Not-I, but with the prior moment—on the order of feeling. In fact in the *Grundriß des Eigentümlichen der Wissenschaftslehre* Fichte is quite clear that the *Anstoß* is to be associated with the primordial self, as a dynamic self-feeling. Fichte describes how the *Anstoß* differs from the idea of the Not-I as follows: "Feeling is the most primordial interaction of the I with itself, and even precedes the Not-I, since, of course a Not-I must be posited to explain feeling" (*FW* 1:369).[39] The Not-

37. Druet, " 'Anstoss' fichtéen," 387.
38. Cited in ibid.
39. Breazeale emphasizes this point. Cited in Breazeale, "Check or Checkmate?" 369.

I is a conceptual construct whose original reference is precisely that of the *Anstoß*. But what then is the *Anstoß*?

Like the original I/eye, the *Anstoß* itself is precisely unknowable. The dialectic of the I/Not-I allows one to delimit conceptually, and thus give figuration to, the *Anstoß*; but it is in itself unknowable. Yet like the discourse of the eye, and the seeing of seeing, one can more closely approximate the *Anstoß*—beyond any conceptual framework, through affection—through a feeling of feeling. The *Anstoß* appears as feeling precisely because our only access to it is through the external influence upon the subject. We become aware of an influence as *feeling*. This influence brings the subject not simply to action, but to an original self-awareness as a subject of that action. Since the very existence and essence of the self is defined by Fichte as action, the *Anstoß* as first mover, must also be understood to bestow existence upon the self. Fichte writes: "The self is merely set in motion by this opponent in order that it may act; without such an external prime mover it would never have acted, and since its existence consists in acting, it would never have existed either" (*FW* 1:279/*SK* 246). The self-feeling self comes to a first self-awareness through an external alteration of its homeostatic system. This alternation is feeling. Self-feeling can only arise through a first external influence. In short, hetero-affection precedes auto-affection. External influence produces feeling, a feeling that is ultimately a self-feeling that is defined as the ego. Fichte confirms the obscurity of the ego's origin in the *Anstoß*: "According to the Wissenschaftslehre . . . the ultimate ground of all reality for the self is an original interaction between the self and something outside of it of which nothing more can be said, save that it must be actively opposed to the self" (*FW* 1:279). The *Anstoß*, like any first mover, is beyond the possibility of explanation. As a consequence, the *Anstoß* stands as a "necessary noumenon" without which "the consciousness of finite creatures" would be "utterly inexplicable" (*FW* 1:279). Fichte explains "that the finite spirit must necessarily posit something outside itself (a thing-in-itself) and yet must recognize, from the other side, that the latter exists only *for it* (as a necessary noumenon)" (*FW* 1:281). The I is able to recognize it as that condition of possibility, its source of Being, which stands as a necessary noumenon. And while this noumenon—whether as *Anstoß*, or eye—is essentially unknowable, Fichte nevertheless attempts to see "the other side" from our finite perspective. As we already noted, the *Anstoß* is recognized as feeling, as a self-affection whose constitution of the self, already implied an intimacy to the other, a hetero-affection that stands as *the* original bond of the social. Yet what is the nature of this influence? Is it causal, like the materialist account of the *Anstoß*, or is its influence affective, or perhaps even aesthetic in nature?

While the idea of self-limitation implies a pure, near solipsistic decision of an already self-conscious being, the idea of an *Anstoß* conveys a brute physical contact and consequent alteration. Although we have been arguing that the *Anstoß* is not actually a physical, causal phenomenon, just how it influences or impacts the other beyond the physical paradigm (in which our own description is entrenched) remains to be seen.

For the ego, the nudge of the *Anstoß* is, in effect, the bestowal of Being. The self perceives its limitation as an act of self-limitation simply because the *Anstoß*, as a precondition of consciousness, stands anterior to consciousness. The *Anstoß* is that ontological element which brings the self into existence, and thus by Fichte's own testament, is that aspect beyond autonomous determination through which the self is opened up to ethical and social relations.

Yet how does one come to relate to this transcendence at the heart of one's own innermost self? How does one come to embrace the equiprimordial other of the self that establishes its conditions of possibility? Or at least, how does one come to give figuration or define the topography of that self which by definition is beyond the consciousness of the self? How does one lay hold of the *Anstoß*, and thus connect at least indirectly to the very source of life and Being?

One way to approach these questions is to come to understand the *Anstoß* as the catalyst for life and Being through the reflexive immediacy of self-feeling. Like the limit of any phenomenology of the visible, one cannot simply turn to see the eye itself as the source of vision; but rather through a subtle (non)reflective immediacy of the seeing of seeing, a certain transparency nevertheless appears in the doubling back in which the invisible itself appears through the transparent veil of the visible.

Yet doesn't our interpretation of such thoroughly metaphysical language belie the decisively important material aspects of the term "affection"? Doesn't the power of affection—that form of self-awareness by which one sees seeing or feels feeling—reside precisely in its capacity to bridge the sensible/intelligible divide? While we will take up the metaphor of the eye again in the context of Fichte's last works in later chapters, for the moment we need to turn our attention to the material aspects of affection as Fichte explains them through his articulation the power of the body.

After all, it is the body that will stand as a relay between sensible and intelligible worlds and as such offer us the means of explaining a noncausal and noncognitive communicative influence. This influence, we suggested, first appeared as the original affection of the *Anstoß*. Like the immediacy of the self-seeing eye, the feeling of feeling constitutes the *ego cogito* through the self-awareness of the self-affective self.[40] The self-feeling, however, is first dependent upon the original affective influence of the *Anstoß* upon the bodily system of sensibility. If we are to grasp the power and form of the communicative influence that lies at the base of the *Aufforderung*, and seems to arise on the order of affectivity or feeling—beyond the paradigm of intersubjectivity—we must turn to further investigate the body as a system of sensibility.

Anstoß and the Bodily System of Sensibility

Self-affection is at heart a hetero-affection because any feeling, including the self-feeling that defines the original self, is only possible through an external influence. Feeling, as defined by Fichte, is an external influence upon a homeostatic system of sensibility. Consequently, our investigation of affection must come to understand the "general system of sensibility" known as the body (*WLnm* 179).

The system of sensibility is itself that which has the capacity for determination, and thus until it is determined by a particular sensation or feeling, remains unnoticed. A determinate feeling is a specific alteration of the system as a whole. It is in this instance that one feels—say hot or cold—that one becomes aware of oneself, such that in feeling cold what one feels in this feeling is simply oneself. At the point of homeostasis nothing is felt at all. The feeling of oneself is a feeling made possible by an alteration of the system. Fichte writes: "I do not feel something; but rather I feel myself" (*WLnm* 180). This self-feeling or self-affection marks the first emergence of a real "I" that becomes ideal through self-intuition. To reiterate, feeling designates the subject's self-relation, while intuition takes this first subjective self-relation itself as its object. What is the thing that first causes an alteration within the system that makes possible an awareness, a self-awareness that we call the self? It is the *Anstoß*, of course.

The system of sensibility is a description of the body, a body whose self-awareness arises out of an interaction with other bodies. The social sphere is

---

40. Emmanuel Levinas, of course, also recognized that the ego was constituted in affectivity. He writes: "Affectivity is the ipseity of the I." See Levinas, *Totality and Infinity*, 117.

from one perspective a dynamic collectivity of bodies. But unlike a theory of billiards, it is not exclusively describable in terms of a causal interaction among bodies. Fichte is well aware that *if* the metaphysical concept of the *Anstoß* is to have any validity within the social sphere, it must also have the capacity to describe an intellectual or immaterial influence, which for nineteenth-century science had much in common with the paradox of action at a distance. In other words, if the *Anstoß* is to function in Fichte's account of the social in the form of the *Aufforderung*, it must able to account for intellectual influence, and thus stand beyond a mere causal account.

The shift from the account of a physical to an intellectual influence was not a major problem for Fichte. Mind and body were merely parts of a double-aspect monism. Fichte explains that the "I as 'pure I,' in its supreme purity, and I as 'body' are entirely the same" (*WLnm* 180). Nevertheless, the fundamental problem of communication, the problem of a transmission across the sensible/intelligible divide, returns to us once again in perhaps its most fundamental form. Fichte is presented with the following dilemma: How can physical action cause an immaterial, intellectual alteration? Or, inversely, and perhaps still more perplexing, how can the immaterial form of spirit cause physical alterations? In other words, we are still grappling with the problem of the relation between the spirit and the letter. And while the notion of "affection" functions as an important third term between spirit and matter, sensible and intelligible worlds, the specific dynamic of its influence nevertheless remains obscure. I would suggest that Fichte's development of the *Anstoß* in the social sphere as an *Aufforderung* reflected his attempt to come to grips with problem of action at a distance from what was originally an exclusively mechanical paradigm. Fichte's attempt to advance beyond an exclusively causal, material understanding of the *Anstoß* ultimately would lead him to articulate the *Aufforderung* on the order of an affection that could explain the problem that had plagued mechanical theories throughout the eighteenth century: the problem of action and influence at a distance. Not surprisingly, we shall see, Fichte's attempt to articulate the dynamic of social interaction across the intellectual/sensible divide and resolve the problem of action at a distance, would lead him to borrow the vocabulary of nineteenth-century sciences.

The influence of human beings upon one another does produce an impact—but not necessarily a physical impact. Our vocabulary is merely grounded by analogy to the physical, mechanically described world. What is truly required for such an explanation is a theory that can explain action at a distance—or in this instance, *Anstoß* at a distance. And while the very notion of the *Aufforderung*, the demand or summons of the other, designates the idea of an influence at a distance, it must be admitted that the very notion of intellectual influence is paradoxical at best. After all, while the idea of the intellectual is precisely immaterial, the notion of influence seems to imply some sort of physical or material force. Not surprisingly, most explanations of action at a distance, even at their most speculative heights, seem to remain trapped within a causal paradigm. For instance, with the theorization of a third term (like ether, magnetism, and/or subtle matter) that would stand as a middle term between spirit and matter, one is able to explain the cause/effect relation merely by positing an invisible matter that mediates these diverse realms. In this sense, there is no true action at a distance.

Yet subtle matter's unique position as a third term between spirit and matter allowed for a change in its own meaning. Let me explain. The power of subtle matter—which was at times interpreted along the lines of a strict causal materialism—gradually came to be understood in more spiritual terms.[41] Subtle matter in this instance gave way to the idea of a "suggestibility" in which the affective rapport established between spirits, conveyed by way of the spoken word, allowed an influence and control beyond the materialist paradigm.[42] This was magnetic psychology's idea of magnetic rapport. Rapport, however, turns out to be predicated upon a unique affective bond—the very affection which described the *Anstoß* and stood as the condition of possibility of the self-feeling self. The idea of suggestibility, which marks both the origin of hypnotism and the psychoanalytic notion of transference, I contend, offers an important insight into understanding the *Anstoß* as a nonmaterial, affective power. After all, the *Anstoß* too, whose action precedes self-determination, functions at a level one designates as the unconscious (*das Unbewußte*).

Like the dynamic of the system of sensibility (in which the feeling of the other through an alternation of the system is what makes self-feeling possible), the affective power of the *Anstoß* is that hetero-affection which makes auto-affection itself possible. That putatively narcissistic monad we call the self then does not begin with self-feeling proper, but rather—in the language of Freud—with a *Gefühlsbindung*, with an affective identification through

41. In fact, the larger history of "subtle matter" shows it to be an idea whose constant development and context-dependent use makes it extremely difficult to pin down.

42. See Ellenberger, *Discovery of the Unconscious,* 110–82.

which self-feeling arises as the result of an initial libidinal tie that is always already an openness to a world. In this sense, I am suggesting that Freud's description (in *Massenpsychologie und Ichanalyse*) of the dependence of the ego upon the other in terms of an affective bond is insightful in trying to understand Fichte's own description of the ground of intersubjective recognition in terms of feeling.[43] The self-feeling self (*Selbstgefühl*), insofar as it arises as a consequence of the *Anstoß*, arises as the effect of an always prior affective bond that stands as the very source of the social.

As we saw at the end of our previous discussion on the spirit and the letter in Chapter 3, communication is fundamentally mimetic. Fichte had reworked Plato's notion of the pure *mimos* such that the noble spirit would imbue his form into the other through a mimetic identification. The noble spirit desires that all others be like himself (SLP 88). It is now clear that the form of this communicative identification is at base affective. Or as Mikkel Borch-Jakobson notes in regard to Freud's work, "Affect as such is identificatory."[44] To speak of "affect" is already to open the problem of the Platonic double. The process of identification which marked the culmination of the communication of spirit as an ideal, an ideal of pure self-reflecting community, must be understood as a mimetic form of identification whose true communicative power lies at base in the affective bond. Since we will not be able to explore in detail the mimetic aspects of influence that define the affective social bond until we turn to Fichte's last works in the final chapters, for now we turn to develop this affective influence through a further articulation of the power of the body in Fichte.

If the human body, according to Fichte, as a pure indeterminacy, stands as a symbol of both freedom and the supersensible, which elicits a universal noncausal, noncognitive influence upon the other, then the body, I will argue, produces in the observer a unique aesthetic—a communicative aesthetic that has much in common with Kant's account of the beautiful (aesthetic judgment) in the Third Critique. Thus, in order to bring the communicative influence of the body into further relief we turn once again to Kant. What Kant's Third Critique offers is an interpretative tool for understanding the aesthetic nature of influence involved in both the *Aufforderung* and Fichte's concrete explanation of it through the notion of subtle matter.

---

43. See Freud, *Massenpsychology und Ichanalyse.*
44. Borch-Jacobsen, *Emotional Tie*, 5, 73.

6

The Aesthetic of Influence

Since Fichte claimed he was the only true Kantian, I turn to Kant's Third Critique and his account of the aesthetic of the beautiful for help in developing the dynamic at work in the phenomenon of subtle matter as a key term in understanding intersubjective influence and its account of intellectual influence as an aesthetic power. Fichte's account of the ground of intersubjective influence in subtle matter is best articulated through Kant's paradigm of aesthetic judgment.

Such an aesthetic reading fits well—that is, coherently—with my two earlier claims, despite some clear differences from Kant. First, subtle matter stands as a posttranscendental proof and consequently fits Fichte's reformulated model of the transcendental. Now Kant argued that while empirical interest in the beautiful—as a form of sociability—prepares the way for intellectual, cognitive assent, such empirical deductions should not be conflated with transcendental ones (*CJ* s41.4; *GS* 5:297, B140). Still, Kantian aesthetics provided Fichte some substantial interpretative traction in his attempt to make sense of intersubjective influence. The fact that for Fichte the very validity of the transcendental claim depends to a certain extent upon the empirical does not deter Fichte from invoking Kantian-style aesthetics to such ends. It is clear that for Fichte the articulation of the actual grounds of intersubjective influence as a kind of aesthetics of subtle matter is of great consequence. Second, Fichte's critique of reflection, as developed earlier (through reference to Dieter Henrich, for instance), is also shown to be of significance for the practical sphere. That original moment of consciousness I described earlier in the theoretical terms of self-feeling is now articulated in the practical sphere in terms of an aesthetic self-relation, a relation that is not self-reflective, but immediate in nature.[1]

Thus, the earlier critique of self-reflection through self-feeling is explored in the practical realm in our relation to the other through the development of the reciprocal influence between self and other in which the affinity between auto-affection and hetero-affection is revealed in terms of feeling understood "aesthetically." Like any account of the aesthetic, what is at issue is an *aesthesis* or feeling. Here then, Kantian aesthetics is marshaled by Fichte to help give interpretative clarity to his own account of transcendental inter-subjectivity both in the dynamic of subtle matter and the self-feeling that constitutes the very basis of the transcendental subject.

The term "aesthetic" has two central uses in Kant. The first refers to its determinate, cognitive-epistemological deployment in the First Critique; and the second, to its essentially noncognitive deployment in the *Critique of Judgment*. If in the first instance aesthetic sensibility is the subjective contribution of sensibility to cognitive representations, in the second instance, "aesthetic" refers to the sensible feeling of a purely subjective mental state, which is neither concept nor object, but is wholly reflective (in the nonrepresentational sense).

For Kant, aesthetic judgment, as understood in the *Critique of Judgment*, refers to a judgment of taste. Kant's concern with the beautiful has less to do with an empirical object said to be beautiful than with the subjective formal conditions that define one's experience or judgment of the beautiful. Kant

---

1. Since, I will argue, Kant's denotation of reflection in fact refers to an aesthetic defined by a self-immediate awareness, his account of aesthetic judgment is in no way opposed to Fichte's critique of reflection (reflection understood as representation), but in fact, explains universal influence on the order of feeling and, in so doing, goes some distance in helping us to understand the power of the percept (*Gestalt*) of the human body in Fichte.

sought to describe or outline the development of any possible judgment of taste. His concern with the facultative processes in the judgment of the beautiful led him to develop the notion of a noncognitive reflection that is quite distinct from the self-conscious reflection that Fichte had critiqued earlier. Now, if one recalls, self-conscious reflection described a subject's attempt to turn back and grasp itself as its own object through the power of representation. By contrast, although aesthetic judgment in Kant is also said to be reflective, it is reflective in a radically different sense: it is noncognitive, nondeterminative, and nonrepresentational. Reflective judgment defines aesthetic consciousness, the mental state of the harmony of faculties of the self-feeling subject. It is not an issue of representation. As Kant explains in the Third Critique, aesthetic judgment is not a "sensuous representation of the object," but rather "a sensuous representation of how the condition of the subject is affected by an act of that faculty." Aesthetic consciousness, like self-feeling, is a form of self-immediacy.

Intellectual Influence: The Aesthetics of Subtle Matter

Fichte's references to subtle matter in the *Grundlage des Naturrechts*—however obscure—are clearly an attempt to explain how the *Aufforderung* and *Anerkennung* function empirically. As a consequence, one has little choice in accounting for such descriptions except to conclude that Fichte had seen subtle matter as a possible empirical proof for his transcendental claims. Such an interpretation is further bolstered by the fact that in the last period of the Wissenschaftslehre, in the "Tagebuch über den animalischen Magnetismus" (TaM 70), Fichte quite explicitly assigns this role of "proof" to subtle matter and magnetic psychology.

Fichte's attempt to develop the notion of "intellectual influence" through the discourse of subtle matter in the *Grundlage des Naturrechts* occurs in the fifth theorem, in "The Deduction of the Applicability of the Concept of Right." While this title suggests that Fichte deduces the concrete application of his previously theoretical analyses, such a conclusion is discounted on at least two fronts. First, even Fichte recognized that one could not transcendentally deduce an empirical or concrete particular. And second, as we suggested, the notion that such empirical analyses stand as a mere confirmation of transcendental claims fits both with his reformulation of the notion of the transcendental and his analysis in the "Tagebuch über den animalischen Magnetismus" of subtle matter as an explicit "proof."

While Fichte's references to "subtle matter" (and the theory of "affection" it entails) in the *Grundlage des Naturrechts*—specifically, in the fifth theorem in "The Deduction of the Applicability of the Concept of Right"—seem to support his transcendental deduction of the social, one nevertheless wonders *whether the "support" subtle matter offers does not in fact signal certain fundamental limitations of the transcendental method itself.* If, as we suggested in Chapter 1, "subtle matter" stands as a rival account of social interaction, we must ask what motivation first led Fichte to incorporate this notion within his transcendental deduction of the social. Would the description of subtle matter function to buttress his transcendental account? Or, looking ahead to his work in the "Tagebuch über den animalischen Magnetismus," might not Fichte's references foreshadow an implicit capitulation to another explanatory paradigm? The explanatory demands of the *Aufforderung* seem to lead Fichte unwittingly to the idea of subtle matter. Thus Fichte intends to offer an empirical proof for his account of the *Aufforderung* and *Anerkennung*. And further, it would seem as if this empirical component is somehow necessary to the very nature of recognition. After all, Fichte does not merely deduce the hypothetical necessity of recognition; he asks how it actually occurs. The key we suggested lies in the human body. But how? It is through the notion of subtle matter that Fichte begins to develop the idea of a noncausal influence, an influence by which intelligible beings can affect one another by means of their own higher meaning (*höhere Sinn*) inherent in the percept of the human form.

With the idea of subtle matter Fichte offers a theory of intellectual influence that appears to operate outside the paradigm of rational, cognitive communication. Subtle matter functions as the key medium in the communicative exchange known as recognition. Recognition is possible for Fichte because the human figure (*Gestalt*) or perhaps even the human face or look (*Gesicht*) compels one to recognition by virtue of its very being. Fichte explains: "A human body must, in its quietude, without any act, produce an influence: subtle matter must here in this influence be posited, that subtle matter becomes modified through the purely quiet figure [*bloss ruhende Gestalt*], and consequently this contained modification modifies the higher meaning [*Sinn*] of a possible other rational being" (*FW* 3:75). The human body or figure through its very existence, then, has the power of conveying a higher meaning or influence to another rational being. Yet Fichte does not stop with the explanation that recognition takes place by virtue of the human form. He further details how the human figure or face produces an intellectual influence—through the medium of a subtle matter—that compels us to recognition.

As Fichte's description of influence, however, relies more and more on a material and quasi-material paradigm exemplified in the notion of "subtle matter," so too must the communicative form come to take part in the more bodily conditioned communication of affection that distances itself from the ideals of a strict transcendental inquiry. Thus, while the summons that made recognition possible is said *not* to operate like the principle of material causality (*FW* 3:36), it is nonetheless curiously described not simply as an influence, but as a material with the capacity for influence. The notion of influence would seem to require some sort of material cause, some matter, no matter how subtle. Fichte writes: "The summons is the *material of influence* [*die Materie des Wirkungs*] and a free influence [*freie Wirksamkeit*] of rational beings" (*FW* 3:36; my italics). It is my contention then that, read in the context of the problem of "subtle matter," Fichte's description of the summons in terms of a "material influence" is no mere turn of phrase, but the result of the extreme difficulty of trying to describe a form of intersubjective influence that exhibits the conflicting properties of something at the limits of the ideal and material worlds. Fichte needs to describe an idea of influence that is free from the necessities of causality and thus preserves a spiritual and rational autonomy, despite its emergence from a material paradigm.

It is no surprise, then, that in the *Grundlage des Naturrechts* Fichte begins to develop the idea of subtle matter in sections 4, 5, and 6 in "The Deduction of the Applicability of the Concept of Right" in which he makes the transition from an intellectual to a bodily account of intersubjective influence. Here, in his explanation of the role of the body in intersubjectivity, he intends to further develop this idea of a noncausal, or quasi-material, influence. Like the first theorem, which posits the self as free, followed by the second, which insists on the role of the other (*FW* 3:17–24); the fourth and fifth theorems mirror this same movement except in the material register of the body (*FW* 3:57–62). In this section Fichte is making the broader claim that the body itself is a necessary condition for the consciousness of right (*Rechtsbewußtsein*). In brief, he is arguing that the intuition of one's own freedom by the pure will is possible only through the body, and that the inference of this freedom within another is the fundamental moment of the consciousness of right. Yet if, as Fichte argues in the *Wissenschaftslehre Nova*

Methodo, the body is the representation of freedom in the sensible world (*WLnm* 320), is the other's freedom merely transcendentally deduced, or does intersubjective recognition also involve something on the order of a quasi-material influence or affection? After all, why speak of bodies at all?

Fichte states in the fourth theorem, a "rational being cannot posit itself as an effective individual without ascribing to itself a material body through which it determines itself" (*FW* 3:56). In other words, the body, as a material manifestation of the pure will, has causal force in the world. The body is the will as act (*WLnm* 327). Here, Fichte seems to affirm a double-aspect monism in which body and will are different manifestations of the same obscure power. Yet even Fichte's insistence that "my body and I—my mind and I . . . mean the same thing" (*WLnm* 321) does not seem to move us beyond the solipsism in which, like Descartes and even the Cartesianism of Husserl, we simply infer or project our own properties upon another.[2] Unless, of course, as any double-aspect monism would require, intellectual communication can also be described in the register of bodily communication—that is, as affection.

As we already noted, the problem facing Fichte, like Kant before him, was how an individual subject to material influence could be conceived at the same time as free. While Kant was able to posit a noumenal realm to resolve this difficulty, Fichte's inclination toward a dual-aspect monism makes his response more complex. Such a problem forces Fichte to reconfigure the traditional psyche/soma problem through a new vocabulary of higher (*höhere*) and lower (*niedere*) organs, which find efficacy in subtle and gross matter respectively. In one sense Fichte has already solved this dilemma, insofar as subjects understand that self-limitation is the transcendental condition of freedom, and therefore freely choose to limit themselves—and thus although determined, as self-determined, they remain free. Yet, in another sense, the intimacy of "body" and "will" allows the problem of material determination to remain a threat to Fichte. Like Descartes's rather improbable solution to the mind/body problem in the pineal gland, Fichte needs to address how nonverbal or noncognitive communication would be possible between two beings across the divide of the material world. While we have already begun to outline this schism, Fichte's articulation of a subtle and gross matter, I believe, is the culmination of his attempt to work out this rather complex problem within the *Grundlage des Naturrechts*.

2. Ferry, *Political Philosophy 2*, 154.

Affective Rapport: Subtle Matter and the Aesthetics of Imitation

The theory of subtle matter offered Fichte a means of explaining communication between material and intellectual worlds and thus allowed the dynamic of intersubjective recognition to escape the problem of solipsism, by conceiving recognition on the order of an "affective" rapport. Yet while we have noted that this communicative rapport between rational beings is not one of rational discourse, we have yet to explain just how this communication takes place. In this section we will begin to articulate Fichte's description of this unique form of communication. Here Fichte will describe this communication across the spiritual/material divide through a process of imitation or copying (*Nachbildung, Nachahmung*), made possible by the medium of subtle matter.

Since Fichte conceives gross matter—quite like the traditional philosophical notion of substance—to operate and influence through a material causality, his description is rather clear and to the point. He writes: "As the very same condition I must posit outside of myself a viscous durable material capable of resisting the free movement of my body, and so the sense world [*Sinnenwelt*] is further determined through the further determination of my body" (*FW* 3:68–69). At the level of material causality, body and world reciprocally determine and influence each other. What remains at issue, however, is the rather difficult and obscure term of "subtle matter." The ground of influence in the deduction of right is the inference of another rational being. The mere material impact of another body, we suggested, is not enough: there must be a higher meaning or sense (*Sinn*)—like that conveyed in the *Aufforderung*. And it is with this assumption of a higher meaning that the body of the subject is truly modified by the other (*FW* 3:69). In fact, subtle matter, I am arguing, is a quasi-empirical explanation of that obscure communicative dynamic Fichte calls the *Aufforderung* (and consequent *Anerkennung*). This is the moment of recognition in which a higher meaning is conveyed from one human being to another—a communicative encounter whose higher meaning constitutes recognition as such, and therefore allows us to differentiate it, for instance, from an encounter with a plant, animal, or mineral. Without this higher meaning a person could not be grasped as rational or human and, in turn, as free. Fichte affirms: "In the described influence the subject's organ [the body] becomes actually modified through the person outside of him. Now this happened neither through an immediate bodily touch of this person, nor as conveyed by viscous [*gross*] matter; since then . . .

the subject would not then perceive itself as free" (*FW* 3:69–70). It is clear then that freedom can be neither assumed nor inferred through a material influence conveyed by what we have been calling gross matter. Communication does not take place at the level of material causality. What Fichte has prepared us for is a type of influence that is not merely part of the substantial causal chain, but a peculiar type of matter that arises from the meaningful body in its totality—a subtle matter.

This influence by which the material passivity of one being conveys a higher, intellectual meaning to another proceeds through the dynamic of imitation. And subtle matter forms the key medium for this fundamental moment of recognition. For Fichte, recognition between spiritual beings is possible through an essential mode of relating between the higher and lower organs by means of a process of copying or imitating (*Nachbildung*) in which the higher organ copies the determined movement (*bestimmte Bewegung*) of the lower.[3] It is this mirroring between spiritual and material worlds that makes communication possible. It is a communication, or order of affection, that takes place through the pure will (*FW* 3:70). This structure of influence, however, applies not only to higher and lower organs in the same being, it also begins to explain how intersubjective recognition would function on the order of subtle matter or, in other words, how the image or percept of the other empirically produces a recognition of right.

As I noted earlier, certainly other rational beings can be deduced, but how does one know when he or she has encountered one? The answer lies here, in the notion of imitating, in which one rational being is able to communicate a recognition of the other through a *mimetic* rapport. Or, in Fichte's abstract language, one might say the following: the subtle matter of the higher organ expresses its capacity to imitate the determined movement of the lower. Fichte explains: "In that case the higher organ is modified through a determined form of the more subtle matter, and held; and so if the person is to perceive, he must suppress the movement of the lower organ, insofar as it relates itself to that part of the higher. But nonetheless even though inwardly it must imitate, in the same organ the determined movement that it would have to make if it itself were to generate the determined given modification of the higher organ" (*FW* 3:70–71). The higher organ, the intellectual aspect, makes a copy of the lower, the material aspect; and this copying is made

3. Or perhaps it's the reverse. Fichte's footnote shows that he himself is unclear about this dynamic. See *FW* 370–71.

possible through the medium of subtle matter. And it is by restraining the lower organs that subtle matter receives an impression and makes an inner copying of the higher organs possible. It is worth noting that there is a striking affinity (and as we are arguing, much more than a mere affinity) between *this* description of restraint and the one present in the dynamic of the *Aufforderung*. Communication in each instance is the effect of restraint. Here there is a restraint of the lower organs vis-à-vis the higher, which makes copying, and thus communication possible, while in the *Aufforderung* the summoner restricts his or her bodily freedom, thereby communicating with, and in effect, summoning the other. This similar dynamic is, of course, no accident. To reiterate, subtle matter stands as a description of the intimate empirical workings of the *Aufforderung*.

Still, this rather abstract account of how the material realm is able to communicate with the spiritual world through the medium, or middle term, of subtle matter has yet to settle the problem of intersubjective recognition. Fichte addresses the problem of human reciprocity and explains recognition through the dynamic of perception and the singularity of the face (*Gesicht*).[4] It is the human figure and face that produces the impression in the aesthetic act of copying. And it is subtle matter which is the medium of this imitative *aesthesis*. Fichte writes: "If a figure [*Gestalt*] in space becomes perceived through sight [*Gesicht*], and becomes inwardly, and very quickly . . . the feeling of the object [*Gefühl des Gegenstandes*], that is, the pressure . . . which is imitated, it produces the figure through an aesthetic art [*um durch Plastik diese Gestalt hervorzubringen*], but the impression in the eye is held as the schema of this imitation [*aber der Eindruch im Auge, wird als Schema dieser Nachahmung*]" (*FW* 3:71–72). Fichte is explaining the capacity for one rational being to be "affected" by another. It needs to be emphasized, however, that even this intimate description of the dynamics of subtle matter seems to be dependent upon an aesthetic description. The dynamic of subtle matter then does not exclude our earlier aesthetic explanation, but is itself an "aesthetic art." The body functions as a higher meaning (*als höherer Sinn*) through this aesthetic of imitation, or in Fichte's words, "*durch Plastik.*"

The phenomena of subtle matter must be understood as aesthetic. After all, it is the perception of another that produces an impression within us through a subtle matter, which, according Fichte, conveys a higher meaning

---

4. The German word *Gesicht* is able to present both the subjective and objective aspects of perception simultaneously because it means both "sight" and "face."

(*FW* 3:71–72). The perception of a human body by another rational being is totally unlike the perception of a mere object, since it conveys a higher meaning in and of itself—or at least that we unavoidably infer that meaning. It has the capacity to affect us through a subtle matter by virtue of its mere existence. Any form of copying, as an aesthetic act, seems to give our own body a higher meaning. But one would wonder whether the copying of one being by another, as a recognition of its rational capacity, is a privileged sort of imitation, or aesthetic. After all, Fichte's central example in this section is the imitation of one individual by another in the dynamic of listening and speaking (*FW* 3:71–72). Not surprisingly, Fichte intends that this imitation that is expressed when one inwardly copies and speaks what one hears be not a mere mime, but an active production. Whether this is actually an active production—which Fichte claims—or merely a process of imitation—has yet to be critically assessed.

The critical proof that the dynamic of subtle matter is at base a peculiar aesthetic explanation, is that, like any art (*techné*), it—that is to say, the very truth of art—is plagued by the threat of imitation. I am, of course, alluding to our earlier discussions (in Chapter 1) of Fichte, Plato, and the problem of mimesis. In other words, although Fichte insists that the copying (*Nachbildung*) present in his theory of communication be a productive imitation, not reproductive, the analogy between pedagogical and magnetic rapport perhaps suggests otherwise.[5] If Fichte's theory of communication proves to be imitative and reproductive, despite his assertions to the contrary, it will have profound repercussions for our understanding of intersubjectivity and the communication that grounds such recognition.

Through a brief look forward to Fichte's more extended engagement with the notion of subtle matter in his later work, the "Tagebuch über den animalischen Magnetismus" (1813), it is clear that Fichte later conceived "subtle matter" as a dangerous imitative power of reproduction. If subtle matter, however, is *not* productive, but—as Fichte later conceded—reproductive, in what way then could one come to understand this dynamic of intersubjectivity which was grounded in a communicative recognition, whose medium in

---

5. If Fichte intends this use of copying to be productive, not reproductive, his use of the term *Nachbildung,* for copying, is perplexing to say the least. After all, he goes to great lengths in works like the *Wissenschaftslehre* of 1794 and "Concerning the Difference Between the Spirit and the Letter Within Philosophy" to distinguish between the productive and reproductive imagination in order to clarify the manner in which *Nachbildung* is an inferior and derivative form of the productive spirit of the imagination.

subtle matter showed the dynamic of social relations to be fundamentally mimetic? In other words, if—as we have been arguing—the communicative exchange known as recognition, is made possible by the medium of subtle matter, and the character of this exchange is a copying that is wholly imitative, then Fichte's entire transcendental deduction of right is put in jeopardy by the more "affective" account of the social that the aesthetic of "subtle matter" represents.

A Phenomenology of the Body: Freedom and the Aesthetic of Communication

The foregoing explanation of subtle matter is admittedly obscure. As we noted earlier (in Chapter 3), Fichte accepts at the outset that a spirit to spirit communication is impossible. Therefore, intellectual influence must somehow cross a material divide. Fichte's language of "*Nachahmung,*" and "*durch Plastik,*" and so on, suggests that this intellectual influence, influence at a distance, is some sort of aesthetic power. Since Fichte's description of this influence, while intriguing, remains deeply mysterious without further interpretation, I would like to try to illuminate further the dynamic of subtle matter in Fichte's *Grundlage des Naturrechts* by developing his understanding of aesthetic influence. I will do so by showing how his account of the human body in terms of freedom and an aesthetic of the beautiful is very much in accord with Kant's account of the beautiful in the Third Critique.

Since Fichte imagined that he alone embodied the true spirit of Kantianism, it makes some sense (as one of several possible interpretative approaches) to try to understand Fichte's claims about subtle matter within the context of Kant's work. In the same section in which Fichte develops an account of intersubjective influence through subtle matter, he also invokes what I understand to be a Kantian-style aesthetics in which he outlines the power of the percept of the human body as an empirical sign of freedom.[6] Now in the same way that subtle matter would stand as an explanatory power that would attempt to mediate the divergent realms of causal and intellectual influence, Fichte postulated the idea of the human body (and its influence) as an empir-

---

6. Ferry emphasizes Fichte's understanding of the body as an empirical sign of freedom in his work. Ferry's interpretation of Fichte's relation to Kant's aesthetics was crucial in shaping my own understanding of Fichte's notion of "subtle matter" within the framework of Kantian aesthetics. Ferry, *Political Philosophy 2*, 166.

ical sign of freedom that would somehow help resolve the third antinomy (causality and freewill) by deepening Kant's account of how moral freedom was possible in the phenomenal world.

While I am in no way offering the definitive interpretation of Fichte's account of subtle matter, I believe that the attempt to understand Fichte's description of the human form as an empirical sign of freedom (that exerts an aesthetic influence) nevertheless goes some distance in explaining the dynamic of influence exerted by subtle matter as the very ground of this intersubjective *Aufforderung* and *Anerkennung*.

As I will ultimately argue, I believe Fichte's description of subtle matter and his account of its influence through the human form (as an empirical sign of freedom) must be read in the context of Kant's aesthetics. Specifically, I contend that Fichte's description of the power of the influence of the human form in intersubjectivity is a radicalization of Kant's account of the empirical basis of the judgment of taste in the *Critique of Judgment* in which an empirical interest in the beautiful prepares the way for rational accord (cognitive agreement) (*CJ* s41).

In order for Fichte's account of subtle matter to make any sense—particularly in light of its reinterpretation of the *Aufforderung* and *Anerkennung*—it is necessary to show how our intersubjective relation to the other is more than merely the sort of intellectual construct that the transcendental deduction of the other represents. I will offer such an interpretation through an exploration of Fichte's account of the human body as an empirical sign of freedom.

First (as we have been arguing), Fichte does not merely deduce the human body of the other. But rather, by emphasizing our immediate perceptive relation to the other (anterior to any intellectual construct), Fichte can be understood to be carrying out a phenomenology of the body.[7] Fichte undertakes a phenomenology of the body insofar he descriptively accounts for its powers and activities beyond any predetermined concept. And it is this analysis of the immediate perceptive grounds of recognition that should give us further insight into Fichte's account of subtle matter.

Fichte offers an account of the human body as an empirical sign of freedom in the same section he explores the notion of subtle matter (*Grundlage des Naturrechts*, sec. 5). His search for an empirical sign of freedom must be understood in the context of the Kantian problem of how moral freedom is

7. Ibid.

possible in the phenomenal world. Of course, another way of asking this is "How is free will possible in a world determined by causality?" Kant recognized the difficulty of such a problem in his explanation of the third antinomy in the First Critique. The third antinomy, in effect, asked how free will would be possible in a causally determined world.

Following the lead of scholars like Alex Philonenko and Luc Ferry, I will show that Fichte believed Kant's solution to the third antinomy to be inadequate. Consequently, he attempted to resolve the third antinomy though the antinomy of teleological judgment.[8] Such a suggestion seems to make sense, since even Kant makes clear that a possible resolution to the third antinomy lies in the *Critique of Judgment's* antinomy of teleological judgment (*CJ* 265–301; *GS* 5:385–415). While we have already briefly explained the third antinomy, we now turn to explore the antinomy of teleological judgment.

In the *Critique of Pure Reason*, Kant argues that the world is not mechanism alone, but from the perspective of the finite transcendental subject, it is penetrated and directed by a purposiveness. Kant is *not* arguing that nature is intrinsically purposive, but rather, as finite beings, there stands a regulative ideal of reason that leads us to the idea of a conceptual totality, to the idea of a purposiveness by which we are able to attribute a coherent unified end to the varied causal mechanisms of an organism and, by extension, to nature as a whole. Teleological judgments then, are subjective, reflective judgments. Reflective and cognitive teleological judgments make use of reason's determinate concept of purpose, but apply it only subjectively. Here, nature is attributed a subjective purposiveness. Consequently, a teleological judgment (as a reflective, rather than determinative judgment) is best understood according to Kant as "a maxim imposed . . . by reason" (*GS* 5:398; *CJ* 280).

Just as the First Critique's critical assessment of metaphysics reined in our speculative flights, but nevertheless made room for practical reason, so too must the mechanistic world be understood to be breached, and in effect supported, by a final cause, by a supersensible substrate associated with freedom. It is nature's capacity to produce organisms—like the human body, in Fichte—that leads to the idea of the supersensible (*GS* 5:381). In other words, the Third Critique, and in particular the critique of teleological judgment, is able to link the First and Second Critiques by showing that the causal chain—and the potential infinite regress—of the First Critique must end in some indeterminate supersensible substrate (a final cause), which nev-

8. Ibid.

ertheless stands as a support to the determinate supersensible (God), which practical philosophy has simply accepted on rational faith (*GS* 5:456; *CJ* 346–47).[9]

In the context of teleological judgment, then, the third antinomy of freedom and causality strives toward a tentative resolution in which the theoretical claims of causality ultimately point toward the supersensible realm of freedom (which had formerly only defined the practical). Insofar as we are not infinite, but finite beings, we cannot rule out freedom or purposiveness in nature. The theoretical postulation of an indeterminate supersensible in teleological judgment readily leads to a moral teleology (*GS* 5:455; *CJ* 345–46) in which the theoretical demand for a final cause is united with the moral concept of freedom.[10]

Now the methodological advantage of resolving the third antinomy through the antinomy of teleological judgment is that by explaining how the notion of finality coexists with the causal world, the antinomy of teleological judgment suggests that another account of the world—perhaps one of freedom—is possible (and not necessarily or truly antithetical). When one asks of the status or reality of the antinomy per se, it is an aesthetic reading that suggests that the antinomy is perhaps more perceived than real. For instance, to what extent can the principle of causality be thought of as a principle of reflective judgment?

Insofar as the principle of causality, like any transcendental condition, is constitutive of experience, such a principle stands as a very model of determinative judgment.[11] Nevertheless, the principle of causality, as a transcendental category, is wholly "regulative" with regard to experience. It does not instantiate "real" existence. As Luc Ferry explains: "The principle of causality is determinative at the level of possible experience and reflective at the level of real experience."[12] The point here is that the idea of finality is invoked as a reflective judgment for at least two reasons. First because an explanation by means of mere mechanism remains, in certain instances, highly improbable; and second, that without such reflective judgments we would be compelled to posit the epistemic reality of objects like God and the Thing-in-itself.[13]

9. My own interpretation of Kant is in part inspired by Werner Pluhar's excellent introduction to his translation. See *CJ*, lxii–lxiii.
10. Ibid.
11. Ferry, *Political Philosophy 2*, 154.
12. Ibid, 155.
13. Ibid.

In the *Vocation of the Scholar* Fichte is at pains to distinguish an organism, as a unity organized by the principle of finality, from mere causal mechanism. He explains: "The questions remains: how to distinguish an effect produced in experience by necessity from a similar effect produced through freedom" (*FW* 4:304).[14] What he suggests is needed "is another distinguishing mark." This new distinction would be between organization and life. To be fair, these three levels or distinctions—mechanism, organization, and life—were already present in Kant. As Ferry maintains, Fichte's true innovation is in offering an empirical criterion for life.[15] Thus, while Kant defines life in the *Critique of Practical Reason* as the "faculty a being possesses of acting according to its representations," Fichte attempts to describe the empirical criterion for life in the phenomenal world.[16] He does so through the notion of the articulated body.

Fichte can be understood to resolve Kant's antinomy of teleological judgment by developing the notion of the articulated body as an instance of free finality in the phenomenal world.[17] Fichte shows that the body stands as an empirical sign of freedom, of free finality distinctly different from both the natural finality of an organized being and the causality of mere mechanism.

The Articulated Body: An Empirical Sign of Freedom

The issue we are centrally concerned with, one recalls, was the relation between Fichte's references to subtle matter and the nature of his transcendental deduction of the *Aufforderung* and *Anerkennung*. The notion of subtle matter first suggested that a recognition of the other and the other's influence upon me was not strictly explainable in terms of a transcendental deduction. And it is this question that leads us to Fichte's notion of the articulated body. Speaking of the percept of the human form, Fichte asks, "When you saw this figure [*Gestalt*] you had to take it necessarily as the representation of a reasoning being in the sensible world, if you are a reasoning being. How is this possible?" (*FW* 3:80–81). In other words, how do we come to recognize another being as a rational being?

While one might be tempted to think of the other as a mere necessary

---

14. Cited in ibid., 164.
15. Ibid.
16. Ibid.
17. Ibid., 167.

condition of possibility, and thus an intellectual construct, it should be emphasized that here Fichte is clearly dealing with the phenomenon of immediate perception, a perception which makes the consequent conceptualization of the other possible. Since for Fichte the body of the other is not reducible to any concept, the perception of the body of the other comes to stand as a sign of freedom. Fichte writes: "I cannot conceptualize the appearance of the human body, except through the presumption that it is a reasoning being, . . . I cannot, in collecting the parts of its manifestation, come to a standstill before I have arrived at the point at which I am freed to think of it as the body of a reasoning being" (*FW* 3:77).[18] In fact, Fichte's emphasis that man alone is "originally nothing at all" (*FW* 3:80) is that culminating moment of the articulated body's persistent rejection of the concept.

The body's refusal to be subsumed by the concept, in fact, is the very ground of its immediacy. Thus, when Fichte asserts that "the articulated human body is sense" (*FW* 3:65), it is the same body as sense that stands as a figure of nothingness, pure determinability, and thus freedom. And it is through these two seemingly antithetical aspects of sense and freedom that Fichte begins to delimit the articulated human body as an empirical sign of freedom.

While humanity is easily distinguishable from mere organization—like that of plants—what is required is a distinction that separates us from other animals, from other articulated bodies. At this point, Fichte is largely following the Kantian definition of living beings, as beings able to act according to representations (and having articulated bodies). What Fichte adds are the qualifying terms of determined and undetermined free movement (*bestimmte und unbestimmte freie Bewegung*) (*FW* 3:79). Such a qualification is made in order to explain that while animals move by representations, and are largely a function of instinct, and thus *determined*, humanity, by contrast, moves according to representations that are the product of free will, and are thus *undetermined*. In this sense humanity stands as pure possibility. Fichte explains:

> Its articulation . . . must be incomprehensible in any determined conception. . . . There must not be a determinedness [*Bestimmtheit*] of articulation, but an infinite determinability of articulation; not a development, but a developability. . . . A rational observer of the

18. Cited in ibid.

> human body can unite its parts in no conception except in the con-
> ception of a rational being like himself, or in the conception of
> freedom given to him in his self-consciousness. He must subsume
> the conception of his own self to his contemplation of that other
> human body, because that body expresses no conception of its own
> . . . every animal is what it is; man alone is originally nothing at all.
> (*FW* 3:80)[19]

It is this visible instance of the human body and its pure possibility in "undetermined free movement" that would seem to offer an empirical sign of freedom, and thus point to a supreme realm of moral freedom. It is in this sense
that Fichte attempts to complete Kant's *Critique of Judgment*. He offers a
justification or criterion for thinking the purposiveness or finality of humanity within the natural world through an analysis of the undetermined free
movement of the articulated body. And it is this material body that stands as
a phenomenal sign or symbol of freedom.

Since the ego is not a particular essence, but is defined by act, the essence
of humanity is precisely a nothingness, a *tabula rasa,* that makes possible
Fichte's understanding of humanity as a power of pure possibility. Fichte
writes: "Every animal is what it is, man alone is originally nothing at all;
what he must be, he must become [*was er sein soll, muss er werden*] . . .
malleability, as such, is the character of humanity [*Bildsamkeit, als solches, ist
der Character der Mencheit*]" (*FW* 3:79–80). Malleability, then, is a description, not only of the character of humanity, but also of the *mode* of humanity's indeterminate freedom. Humanity's "mode" is its malleability, and this
Bildsamkeit, as *das Bild,* also expresses its capacity to stand as a transparency
to what is transcendent to it.[20] Humanity's fundamental capacity or vocation
then is to stand as an transparent image to that which is transcendent to it.

The Body as Symbol and Empirical Sign of Freedom

While an array of mechanical causes can be unified through the attribution
of a specific purpose we designate as an organism, it is the *human* body, we

19. Ibid., 169.

20. This description of humanity in terms of a malleable image is quite suggestive in view of his
later articulation of the imagination as a metaphysics of imaging in the last period through the notion
of the individual as a transparent image or conduit of the absolute.

recall, according to Fichte, which—since it cannot be attributed a specified purpose—is precisely itself a sign of the supersensible. Kant's resolution of the third antinomy through teleological judgment allows us to understand how the body uniquely stands at the juncture between phenomenal and noumenal worlds. If we are to understand how the body stands as a symbol of freedom in its indeterminacy as the distinguishing character of humanity, Kant's account of the beautiful as the "distinguishing property of man," leads one to surmise that an investigation of the power of the human body in intersubjectivity would further demand an account of aesthetic judgment.

Following Kant's resolution of the third antinomy (free will versus determinism) through the antinomy of teleological judgment (mechanism versus purposiveness), we can see how the body for Fichte, as an indeterminate organism, stands both within and without the order of causal mechanism. It stands as a symbol of the supersensible (freedom) within the causal, sensible world. Nevertheless, while our discussion of teleological judgment in Kant helped clarify how the body stood for Fichte as a symbol of freedom, the actual empirical dynamics of *Anerkennung* and *Aufforderung* still remain obscure. If we are to understand this fundamental form of communication in Fichte—which defines the condition of the possibility of the social, and in turn subjectivity—we need to grasp how one functions as an influence on the other. The problem of action at a distance is still with us. After all, if freedom is possible, the *Anstoß* must function beyond the order of causal mechanism.

Beauty, in the instance of Kant, and the human body, in the instance of Fichte, each carry the same signifying weight as a symbol. In each case the symbol signifies freedom, morality, and the supersensible. This symbol of beauty and/or the body was made possible by the rapprochement of sensible and intelligible worlds through the explanation of teleological judgment. Kant's explanation of the relation between the two forms of reflective judgment (known as teleological and aesthetic judgment respectively), brings into relief Fichte's understanding of the role of the body both as the mediator between sensible and intelligible worlds, and as such, as the fundamental link of the sociopolitical body, whose formative power stands well beyond its association or analogy to the political body. Since teleological judgment identifies the subjective purposiveness of a given organism in nature, things turn out to be much more complex with Fichte's account of the organism of the human body, which—like Kant's account of beauty—is defined as a purposiveness without purpose. Here we note that the two forms of reflective judgment—the teleological and the aesthetic—cross over into one another. The human body, as a symbol of freedom and the supersensible, is a work of beauty, a purposiveness without purpose.[21] In other words, I am arguing that the key to understanding the body as a power of communicative summons, through its symbolic reference to the supersensible, lies in part in Kant's transition between teleological judgment and the judgment of taste. Kant himself was already profoundly aware of the intimacy of teleological and aesthetic judgment. Kant explains the intersection of these two forms of reflective judgment as follows: "Hence the object is called final, only because its representation is immediately connected with the feeling of pleasure; and this representation itself is an aesthetic representation of the finality" (*GS* 5:189).[22] The human body, as Fichte articulates it, however, makes the intersection of teleological and aesthetic judgment still more clear.

While neither type of reflective judgment (teleological or aesthetic) is determinative (i.e., determine objects), teleological judgment is typically based on the concept of purposiveness *with* purpose, while aesthetic judgment refers to a purposiveness *without* purpose. In other words, while with teleological judgment, through the concept of nature's "subjective purposiveness" (nature's purpose for our subjective judgment) we might attribute a particular purpose to a given organism—like a snake—the aesthetic judgment of the beautiful is wholly without purpose. In Kant's words, "Apart from a relation to the subject's feelings, beauty is nothing by itself" (*GS* 5:218; *CJ* 62–63). By contrast, teleological judgment is a reflective judgment that nevertheless makes reference to a concept. Like the human body in Fichte, though, the purposiveness of the beautiful is without concept. And still more important for Kant's point: the "beautiful is what without a concept, is liked universally" (*GS* 5:219; *CJ* 63–64). The judgment of taste allows Kant to deduce a subjective universality in which one can claim, a priori, a universal assent, a common bond beyond the subsumption of the particular to a universal concept. Kant calls this idea of a subjective universality to which all must assent a priori, a common sense, or *sensus communis*. Through the idea of the *sensus communis* Kant gives us a unique glimpse of the intersection of aesthetics and politics by which the individual can be

---

21. While not all bodies, of course, are empirically beautiful, I am suggesting that for Fichte the recognition of the indeterminateness of the human form, as a symbol of the supersensible, effects within the observer something on the order of what Kant describes as that harmony of the faculties known as aesthetic judgment.

22. Emphasized and cited in Guyer, *Kant and the Claims of Taste*, 68.

shown to relate to the universal of the social whole beyond a determinate concept.

While Kant himself makes clear that the antinomies of teleological and aesthetic judgment are manifestations of the same fundamental antinomy, the antinomy of pure reason, this reciprocity is perhaps most clear in Fichte's articulation of the human body. As a pure nothing, the human body stands as both a material symbol of the supersensible (teleological antinomy), and like Kant's account of beauty (judgment of taste), as an entity characterized by its power to effect an aesthetic judgment. (An aesthetic, we shall see, that stands at the center of our very articulation of the ego as auto-affection.)[23] The body, then, evinces an aesthetic of pure reflection by which one's subjective self-awareness is given universal validity as a *sensus communis,* or in the register of Fichte, as universal—as a fundamental social bond, consecrated through an awareness of one's own freedom, made possible by the percept of the body of the other.

While the idea of an "undetermined free articulation" can be understood to explain (empirically) free will and the freedom of human action, it is nevertheless the means of influence through the percept of the human form with which Fichte is most concerned. In other words, while the free will chooses its own action, this action is nevertheless determined by a freely chosen concept, whereas, by contrast, the indeterminate human form, as a percept of pure possibility, is an aesthetic phenomenon: its image offers only a formal indeterminate purposiveness and, in this sense, functions much like a judgment of the beautiful. Just how the human figure and its indeterminateness effects a judgment of taste, however, can be clarified through Fichte's explanation of the body of the artist in the "Spirit and the Letter of Philosophy." Such a discussion continues our analysis of Kant's notion of "common sense," while returning us to our earlier discussion of how Fichte's critique of reflection is carried over into the practical sphere precisely through a turn to aesthetics and aesthetic reflection.

My theoretical analysis of the ground of self-reflection in self-feeling sought to delimit a self that exists prior to what Dieter Henrich, for instance,

---

23. While I suggested earlier that Kant was aware of the intimacy of aesthetic and teleological judgments, insofar as the judgment of finality was linked to the feeling of pleasure (*GS* 5:189), I am certainly not suggesting that there is an unproblematic relation between the two forms of judgment. In fact, following Philonenko and Ferry in suggesting that Fichte hoped to unify Kant's teleological and aesthetic judgment through the notion of the empirical sign of the human body is not to claim that he succeeded.

referred to as the paradox of reflection. In the same way that Fichte had argued for the transcendental necessity of the other for the ascension to self-consciousness and freedom, I have been attempting to show a primordial ground of the social anterior to reflection in which the "self-other relation" can be understood on the order of feeling in which, here, hetero-affection itself stands as the condition of possibility of auto-affection. By attempting to interpret the discourse of subtle matter and its largely empirical account of the dynamic of intersubjective influence in terms of an aesthetic of influence, I have been suggesting—specifically through reference to Kant—that the concept of self-conscious reflection is preceded (genetically) by the idea of something akin to aesthetic reflection (like that manifest in the judgment of taste). Since what we are grappling with here is the notion of community and communication, we must turn to look at Kant's notion of common sense in still greater detail to see the extent to which the notion of aesthetic reflection can be understood to be communicable (as an intellectual influence) in Fichte.

Common Sense (Sensus Communis)

Common sense (*sensus communis*) is common. It stands as the possibility of a subjective universal claim precisely because this feeling or self-feeling results from "the free play of our cognitive powers," or attunement that occurs universally in every rational being (*GS* 5:238; *CJ* 88). Kant emphasizes that "it is not the pleasure, but it is the universal validity of pleasure" with which he is concerned—yet it is nevertheless feeling which remains the form or conduit of this universal communicability. Through the method of hyperbolic doubt, we recall, Descartes's quest for self-certainty is resolved not in a cognitive claim, but in a subjective universality in which what he could be most sure of was not *what* he felt (tasted, or saw), but merely *that* he was engaged in the activity of feeling. And it was this phenomenological and existential self-feeling that marked the original form of the ego. Here ipseity was achieved in auto-affection. Still, for Kant, the shift from that auto-affective aspect of aesthetic judgment (feeling) to the communicative exchange designated by what we have been calling hetero-affection is far from clear.

To the extent that the *sensus communis* represents the possibility of "the universal communicability of our cognition," much is at stake here for Kant. Yet if for Kant aesthetic judgment offers "proof" of the intersubjective and thus universal (objective) nature of subjective judgments, their origin in aesthetic feeling gives us a further foothold in attempting to understand the affective basis of the social relation. Since aesthetic judgments are nondeterminative, noncognitive judgments, what stands in for the conceptual predicate seems to be precisely feeling (*GS* 5:288). In fact, according to Kant it is the "feeling of pleasure (or displeasure)" which "serves . . . in place of the predicate" (*GS* 5:288). In other words, I, as subject, am feeling myself. But since the very constitution of the ego is defined by the auto-affective feeling which purportedly stands as predicate, what we are dealing with in the phenomenon of aesthetic feeling is a mental state already anterior to the representational structure delimited by the terminology of subject and predicate.

Yet as many commentators recognize, the exact nature of the *sensus communis* is unclear.[24] While the *sensus communis* can be thought of as a principle of judgment, it is nevertheless also a feeling.[25] The difficulty Kant enters into is to explain how a subjective feeling could attain a universal communicability. Kant suggests that the possibility of such a universal communicability lies in feeling. He writes, "If cognitions are to be communicated, then the mental state . . . must also be universally communicable. For this attunement is the subjective condition of [the process of] cognition, and without it cognition . . . could not arise" (*GS* 5:238; *CJ* 88). Affectivity is what makes cognitive communicability possible: it is the mental state or feeling which makes possible the universal assent defined by the *sensus communis*. Our feeling of this attunement, the subjective condition, is what establishes a universal communicability. In other words, if the predicate of this noncognitive judgment is itself feeling—the feeling of pleasure or displeasure—then what is being felt is feeling itself. In more Fichtean terms, it is the feeling of feeling, that immediacy beyond the pleat of reflection—epitomized in the system of sensibility known as the body—that is the ground of the universal communicability Kant refers to as *sensus communis*. Such a claim, however, demands more explanation.

While Kant's account of aesthetic judgment is extremely insightful in helping us understand the dynamic of influence exerted by the human body in the *Aufforderung*, Kant's approach, from the perspective of Fichte, remains too limited: his account of community is not dynamic enough. For Kant, as

24. Donald W. Crawford emphasizes this in his work *Kant's Aesthetic Theory*.
25. While Crawford emphasizes that the *sensus commnis* can be understood as a feeling, Guyer seems less convinced that it can be understood in these terms. For insight into this contentious point, see Crawford, *Kant's Aesthetic Theory*, and Guyer, *Kant and the Claims of Taste*.

we noted, the *sensus communis* defined a common feeling, a common state that could be predicated of others universally. Yet while such a claim makes possible both knowledge and communication by establishing the universality of certain subjective judgments, it nevertheless only explains the possibility of communication, rather than the initial act or dynamic of communication which first constitutes the original moment of social interaction. The difference between Kant's and Fichte's respective approaches might be further brought to relief through the language of "faculties." If for Kant the aesthetic powers of the imagination are ultimately demarcated by the parameters of the understanding, for Fichte the imagination is given no such bounds. Let us look then at the mental state called "aesthetic pleasure" from the perspective of the faculties.

The universality that occurs in aesthetic judgment and produces the feeling called pleasure results from the free play between the imagination and the understanding. The relation between these faculties, however, is more complex than the term "free play" would indicate. Thus, the pleasure one finds in aesthetic judgment is "the imagination's free conformity to the law"—the law of the understanding. For Kant, while a harmony is established between the imagination and understanding, the representational structure of consciousness still demands that the understanding retain its throne.

Fichte's attempt to delineate a pre-representational self led him to radicalize the power of the imagination so that the understanding would no longer be the seat of transcendental spontaneity; it would now be located in the imagination.[26] Thus, if it is in the judgment of the beautiful that aesthetic consciousness (as the harmony of the faculties) arises in Kant, and thus as a *sensus communis* constitutes an intersubjective validity, what is the dynamic of communication that arises in intersubjective recognition for Fichte with the appearance of the human body as a symbol of the supersensible? To what extent are we to speak of an aesthetic consciousness in Fichte, and if so, how? In other words, what is the equivalent of the Kantian harmony of the faculties that, in the instance of Fichte, arises in the process of recognition with the percept of the human body? An answer to such a question, of course, would require a reinterpretation of "common sense" in more explicitly Fichtean terms.

In "The Spirit and the Letter of Philosophy" Fichte engages quite explic-

---

26. Rudolph Makreel emphasizes this point in his article "Fichte's Dialectical Imagination."

itly with the Kantian notion of *sensus communis* (*Gemeinsinn*). He explains his use of the term through the image of the artist. Fichte writes: "What the inspired man [*der Begeisterte*] finds in his breast lies in every human breast, and his capacity is the common capacity [*Gemeinsinn*] of the whole species" (SLP 88). We should note two things. First, that here for Fichte, *Gemeinsinn*, that universal communicative capacity we have been calling *sensus communis* (which appears as aesthetic feeling), is not distinct from our earlier concern with spirit (*Geist*). It is the one who experiences aesthetic feeling that is the *Begeisterte*. And second, that it is the figure (*Gestalt*) of the artist itself—and not any mere beautiful object—which has the capacity to convey this *Gemeinsinn*, and thus signals the significance of the human form for establishing community in the work of Fichte. Ultimately then, as witnessed in the difference between the "Spirit and the Letter of Philosophy" and the *Grundlage des Naturrechts*, Fichte makes the transition from the figure of the artist as a concrete universal who, as a point of access to the absolute, conveys *Gemeinsinn*, to that point in the *Grundlage des Naturrechts* in which the human body of each individual functions as a work of art. As a transparent conduit of the absolute, each human body stands, as the artist, as an incarnate work of beauty.

Of course while one might argue that some bodies are more aesthetically pleasing, more beautiful than others, such an argument misses the point. What Fichte is trying to demonstrate is that it is the percept of the human form itself, the mere recognition of another human being—in its most basic form—which effects that reflective judgment within us called a judgment of taste. Thus, Fichte suggests, it is a consequence of this recognition of another rational being, by means of the percept of the human form, that we experience an aesthetic response that Kant referred to as common sense (*Gemeinsinn*).

So far we have been speaking primarily about how the intellectual influence another material being has upon us can be understood as an aesthetic influence. Yet I would like to briefly consider how we might "aesthetically" relate to our own bodies, precisely in order to make sense of how Fichte's earlier critique of self-reflection could be understood in the realm of the practical. The common term uniting these diverse issues—the issue, for instance, between hetero-affection and auto-affection—is feeling. The articulated body is an active self-feeling body whose "mental state" of aesthetic feeling is not distinguishable from the auto-affection that constituted the ego itself. Yet if this aesthetic is described as a self-feeling, as an auto-aesthetics, in what way can aesthetic judgment then be conceived as a reflective judgment? How can we extol the power of aesthetic judgment as a mode of reflection, if our broader task was precisely to make clear how Fichte's critique of self-reflection extended into the realm of the practical?

Let us be clear. After our lengthy discussion of Fichte's critique of the philosophy of reflection, we are *not* simply eliding his account of intersubjectivity with a form of aesthetic reflection in Kant. To the contrary. Our excursus through Kant was precisely to bring Fichte's critique of philosophical reflection to bear on his account of intersubjectivity in order to flesh out a more primordial social bond, which, like the self-affective subject, grounded its critique through the notion of affectivity. We sought the hetero-affective ground—by reading the *Anstoß* and *Aufforderung* as an aesthetic phenomenon through the percept of the human body—which stood as the condition of possibility of the auto-affective, self-feeling self. Returning to Kant then for a moment, we should realize that the reflective judgment of taste—to the extent that it does not engage in the subsumption of the particular to the universal—is not, I would argue, a philosophy of reflection in the typical sense—or at least in the sense Fichte had argued so vehemently against. In fact, as we shall see in the next chapter, Kant's account of reflective judgment is—far from a philosophy of reflection—much like the feeling of feeling, the thinking of thinking, or the seeing of seeing: it is much closer to what Fichte described in his last works as a pure thinking (*reines Denken*), or what we have been calling self-immediacy.[27]

The body can be understood to be an aesthetic phenomenon in at least two senses. The first is the Kantian sense of the beautiful in which it is a mere percept of the human form which catalyzes a harmonization of the faculties, and thus, a judgment of the beautiful in the perceiver. Now the source of the body's power to effect an aesthetic judgment resides in its capacity to stand as a sign of the supersensible. Now the second and related sense of the body as an aesthetic phenomenon is further developed in Fichte's later work, to the extent the body is understood to actually stand as a conduit to the supersensible. In fact, from about 1799 on, Fichte articulated the role of the body, and the very ethical task of human existence, as one that should aspire to be transparent (*durchsichtig*) to the absolute. Like the artist in "The

---

27. Fichte's notion of *reines Denken* is cited by Wilhelm Wurzer in "Fichte's Paragonal Visibility." Fichte's notion of *reines Denken* is unique. It operates in the following manner. As Michel Henry would say, it is radical immanence or interiority that itself makes transcendence possible.

Spirit and the Letter of Philosophy" that embodied the notion of common sense (*sensus communis*) by standing as a concrete universal, here one aspires to social wholeness by giving oneself over to the absolute in and through total transparency. And it is this new sense of self-immediacy that exceeds the problem of self-conscious reflection through the terminology of *reines Denken*.

Admittedly, like the lure of the beautiful, the percept of the body of the other implies a certain inexplicable fascination. Yet like the power of imitation in Plato, or any true power, this influence, which explains our fundamental social bond, also holds out the possibility for either a utopian ethical relation, or the monstrosity of fascism. In the second instance, influence, as the lure of the beautiful, offers the possibility of a reintegrated social unity through an aestheticization of politics witnessed perhaps in its most monstrous form in German National Socialism. Like the crisis of the Platonic *mimos,* the power of influence clearly holds out both possibilities.

Fichte's work developed significantly in his own lifetime in several respects. First, the self-immediacy we had articulated earlier on the order of "affect" is now read metaphysically as a *reines Denken* in which the displacement of the transcendental subject is so thorough the thought which thinks itself no longer operates as a transcendental insofar as it finds its source in the absolute. It is a *reines Denken* because the subject no longer comes to grasp itself as object; rather, thought thinks itself. Second, to the extent the transcendental deduction of the social in the vocabulary of *Aufforderung* and *Anerkennung* additionally required a posttranscendental proof, the quasi-empirical phenomenon of subtle matter would come to play an important role in Fichte's description of the social. In fact, as we will see in the last two chapters of this book, through an analysis of the "Tagebuch über den animalischen Magentismus" and the *Wissenschaftslehre* of 1813, Fichte continued to develop his account of the social through the polar dependency of philosophical and empirical accounts. With his shift, however, from an explicitly transcendental to a more metaphysical account, the paradigm of magnetic psychology became still more important. Fichte's new metaphysical account of the social in which social unity is achieved though participation in the absolute (by means of transparency) is mirrored in this world in the transparency visible in the phenomenon of magnetic rapport.[28] In brief, what

---

28. One thinks, for instance, of how, in the hypnotic relation, the will of the patient becomes transparent to the will of the hypnotist.

Fichte hopes to find is not a mere empirical proof of his philosophical claims, but what he calls a "Physicirung des Idealismus" (TaM 70): a physical proof or instantiation of the claims of idealism.

As I will detail in the following chapter, what we are witnessing is a double displacement of the transcendental imagination. The faculty of the imagination, once displaced with the dethronement of the transcendental subject in Fichte's shift to a metaphysical account of the social, is deposed again with Fichte's recognition that the imagination itself is capable of being directed by magnetic psychology and its material technique. In the language of the social medium, the creative capacity to produce images, anchored in the transcendental subject, known as the faculty of the imagination, has been externalized in and through material technique, so that the transcendental subject is now less a producer of images than a subject that is largely relegated to the mere reception of images. As I noted in the introduction with Benjamin's account of the decline of "aura" in his essay "The Work of Art in the Age of Mechanical Reproduction," what we have here is an externalization of the power of imaging through material technique. The faculty of the imagination, in part, has been externalized. And while it is in this sense that this analysis of Fichte's encounter with magnetic psychology forms a significant moment in the genealogy of media theory and our own attempt to grasp the power of the technology reproductive media on our own philosophical account of the social and its medium, we have yet to detail what is perhaps the culmination of Fichte's encounter with magnetic psychology in the "Tagebuch über den animalischen Magnetismus."

7

The First Displacement

From Subjectivity to Being

Toward Spirit as Technique

Fichte references the paradigm of subtle matter as an empirical proof for the last Wissenschaftslehre. The problem is that this empirical proof threatens to overshadow that which it was intended to prove: the social ontology of Fichte's metaphysics of imaging. Here if the technique of magnetic rapport overshadows the very metaphysics of imaging it was intended to prove, then Fichte's metaphysics of the imagination, and in short, spirit, may well be relegated to the function of human technique and technology. The crisis of spirit at hand, as Heidegger was well aware, is revealed in technology's attempt to complete metaphysics.[1] Now although Fichte will largely retreat from such conclusions, several points force him to recognize the significance

1. Heidegger, *The Question Concerning Technology*, 32–33.

of technology for a metaphysics of spirit: (1) his awareness of the importance of the mediation of the technology of the letter (writing) for the conveyance of spirit; (2) his demand that philosophy supply an empirical proof, as manifest in both the dual series and his call for a "Physicirung des Idealismus"; (3) and finally, his own acknowledgment and occasional reference to spirit itself as a machine.[2]

The crisis of metaphysics, coincident with idealism's encounter with scientific materialism, aspires to a possible completion and resolution in technology. Such an encounter reaches a visible culmination in Fichte's work in his attempt in the "Tagebuch über den animalischen Magnetismus" to provide a physical proof for his metaphysics of imaging through the technique of magnetic rapport, which, in effect, stood as an external technique or technology of the imagination. Thus, while from a broader perspective this moment in Fichte's work can be understood within the context of a horizon that asks about the possibility of the completion of metaphysics by technology—or even the possible completion of idealism by scientific materialism—and the social consequences of such a resolution, the central thesis of this chapter nevertheless remains committed to the details of Fichte's work—specifically, the significance of the "Tagebuch" for the last period of the Wissenschaftslehre. The very development of Fichte's conception of the Wissenschaftslehre shows that the imagination is not a fixed entity. In fact, I will argue that insofar as the imagination initially arises as a function of the dialectic of striving and resistance—read realistically (as opposed to idealistically) from the perspective of the *Anstoß*—the imagination appears as a function of social praxis, and thus as a product of history.[3]

In view of the "Tagebuch über den animalischen Magnetismus," Fichte's understanding of the imagination can be understood to have undergone a double displacement. This chapter develops the first of these two major transitions. The first shift explores the transition from the transcendental imagination in the first period of the Wissenschaftslehre, understood as a faculty, to the last period of the Wissenschaftslehre's conception of the imagination in terms of a metaphysics of the image in which the subject is no longer the source, but a mere conduit in the process of (divine) imaging. The following chapter, through a continued analysis of the "I," will develop the second

---

2. In *Über Belebung und Erhöherung des reinen Interesse für Wahrheit*, Fichte writes, "Der Mensch. . . . Sein Geist ist eine Maschine. . ." (*FW* 8:343).

3. This interpretation, of course, assumes a certain understanding of the controversial relation between the Wissenschaftslehre and the *Grundlage des Naturrechts*, which we reviewed in Chapter 5.

transition, in which the metaphysical account of the imagination in the last period of the Wissenschaftslehre, as something external to the subject, is read from the perspective of its empirical proof in the "Tagebuch" as a function of human technique. Specifically then, my claim is as follows. It is widely recognized that Fichte's later work represents a displacement of the transcendental imagination in which the creative, generative source of the self now becomes external to it (in the absolute): the transcendental I gives way to the discourse of Being.[4] The second, unacknowledged, and thus more controversial claim, which I am asserting, maintains that a second displacement of the imagination occurs with Fichte's search for an empirical proof of his metaphysical claims in which the metaphysical power of imaging becomes dominated and usurped by its empirical proof in the technique magnetic rapport. Here the second displacement identifies a shift from a metaphysical account of an external imagination to one grounded in material technique.

And although Fichte himself is uneasy with such developments—and even ultimately dismisses magnetic psychology—it is precisely his work with magnetic psychology, before his turn against it, I believe, that is of greatest philosophical interest. Specifically, my analysis of Fichte's attempt to ground idealism through the empirical proof of that quasi-material paradigm, as represented by magnetic psychology, situates Fichte's work within the broader context or genealogy of media theory (understood as the attempt to assess the impact of technology upon that communicative exchange network which defines the social). To be clear how such claims stand as a culmination of this project's developing argument, however, several more minor threads or theses need to be brought together.

Spirit and Intersubjectivity

In the preceding chapters I suggested that Fichte's critique of self-reflection through the idea of self-feeling was also visible in his account of the social in the *Grundlage des Naturrechts*. I explained that intersubjectivity's end of self-consciousness was the effect of the deduction of the other; and like reflection, such an account ultimately rested on the notion of affection or feeling. Through a detailed analysis of subtle matter in the *Grundlage des Naturrechts*

---

4. This interpretation of the imagination in Fichte's last period is held by almost every major scholar of Fichte's later work. See, for instance, Dreschler's *Fichtes Lehre vom Bild,* and Janke's *Fichte: Sein und Reflexion,* as well as the work of Adolf Schurr, Miklos Vetö, Xavier Tillette, and others.

I argued that Fichte himself believed that his transcendental account of the social remained incomplete without explaining one's own actual, affective relation to the other. While I had explained Fichte's interest in subtle matter as an empirical proof for his transcendental analysis in terms of the dual series of the real and the ideal, his explanation can also be understood as part of an effort to resolve the fundamental incompatibility between spirit and intersubjectivity.[5]

The totality of spirit, in principle, is never achievable by means of the dynamic of intersubjectivity. Here, one must understand self-conscious reflection and intersubjectivity as inherently alienated to the extent that, as modes of determinate thinking and reciprocal determination, spirit's ideal of social wholeness stands beyond their reach. Intersubjectivity, modeled on the very schema of reflection and epistemology's subject/object dualism, acknowledges the other in the game of adequation (*adequatio*) as a means to self-consciousness and self-identity that leaves the other at an infinite remove.

This conflict between spirit and intersubjectivity, I believe, is exhibited nowhere more clearly than in Fichte's concept of activity. By defining the "I" beyond any substance, as sheer activity, Fichte overturns an idealism based on categories and shows how consciousness arises from material practices—much like Marx's critique of idealism almost a century later.[6] The problem, however, is that Fichte's initial articulation of activity serves two masters: first, it offers a genetic history of consciousness in terms of the I's phenomenological encounter with the material world as a dialectic between the I and Not-I that cannot thematize a social relation to the other except by means of intersubjectivity; and second, its sense of activity develops a prereflective, pre-determinative, self-feeling self, whose immediacy with the world defines the unity and wholeness that is the hallmark of spirit.[7] After all, Fichte would define Being itself not as an abstract ontological category, but precisely in terms of affect (*FW* 5:498). Still, like Hegel, Fichte in part sees this immediacy as a wholeness not yet achieved by self-consciousness, and it is thus the task of spirit to make this return as a process of recollection. Yet this recollection or return is never possible for Fichte because this dis-

---

5. I discussed this thesis of the incompatibility between spirit and intersubjectivity, held by Theunissen, Habermas, and others, in Chapters 1 and 5.

6. See Rockmore, *Fichte, Marx, and the German Philosophical Tradition*.

7. If it does not represent spirit as *nous*, then perhaps it is more in line with its articulation as *pneuma* (see Chapter 1). It is, in the language of Michel Henry, a hetero-affection that subtends auto-affection.

tance, which is to be overcome in the dynamic activity of striving and resistance, manifest in the antagonism between the I and Not-I, marks the interminable recovery of this self-alienation.[8]

Now while one might rightfully argue that Fichte defines spirit in the first period of the Wissenschaftslehre as a productive imagination that would seem to embody both aspects of activity, in the last period of the Wissenschaftslehre Fichte's radical revision of the imagination away from a transcendentally conceived activity suggests that he was not satisfied with an approach that explained activity in terms of the self's opposition to the world.[9] I would argue that Fichte's recognition of the alienation inherent in his earlier dialectical conception of the imagination, and the self's relation to the other, compelled him to reformulate "activity" in terms of a process of imaging that more readily aligned the imagination with spirit's ideal of social wholeness.

While he in large part achieved this by defining the activity of the I in terms of a near passive imaging in which it stood as a conduit or transparent image of the absolute, the question of practice emerges again in Fichte's search for an empirical proof for his metaphysics. In the first period, intersubjectivity's model of self-reflection would be supplemented in the countervailing aspiration for spirit in the instance of the immediacy of the social bond witnessed in the discourse of subtle matter. And while the paradigm of intersubjectivity would assure the impossibility of spirit, the articulation of consciousness as a social consciousness that arose from the I's encounter with the world, nevertheless, articulated an important phenomenology of material practice.[10] The problem, of course, is that while thinkers like Marx would certainly appreciate Fichte's development of consciousness from the I's material encounter with the world as a form of material practice, to offer subtle matter as ground, proof, or material practice of intersubjectivity is nevertheless to conflate two different orders: it is, in effect, to offer a material practice of social wholeness or spirit as proof of intersubjectivity.

While it is the imagination's very task of reconciliation which leads Fichte to define spirit in terms of the imagination in the first period of the Wissen-

8. This is one of the major sources of the incompatibility of intersubjectivity and spirit. And even some Hegel commentators, like Alexander Kojève and Tom Rockmore, believe that spirit never achieves its full completion: its activity remains interminable.

9. In fact, Fichte began to articulate a new concept of imagination in what is referred to as his second period, after 1800.

10. For an overview of Fichte's relation to phenomenology, see the excellent collection *Fichte and Phenomenology*, or my article in particular, "Reduction or Revelation? Fichte and the Question of Phenomenology."

schaftslehre, I would argue that the intensely dialectical structure of the imagination is modeled on the very antithesis inherent in the I's striving and limitation, self-conscious reflection, and intersubjectivity, in such a way that it is—in reality—not compatible with spirit. Or at least it is a process of spirit that is ineffective to the extent it can never achieve its end or ideal of social unity. And it is Fichte's quest for spirit that leads him to move beyond the paradigm of intersubjectivity and self-conscious reflection, through a radical rearticulation of the imagination, in the last period of theWissenschaftslehre.

In the last period, Fichte's development of activity itself in the language of imaging establishes a paradigm of social unity that is wholly in accord with spirit. The problem of the relation between philosophy and its empirical proof here, however, is different. If with intersubjectivity Fichte established a transcendental deduction of the social in line with his account of the imagination, I argue that his analysis of the actual practice of the intersubjective *Aufforderung* and *Anerkennung* led to a discourse of subtle matter whose concrete immediacy suggests a form of spirit profoundly at odds with his initial conception of the imagination as a work of spirit. Thus, like his earlier concern with the dual series of the real and the ideal, Fichte sought an empirical or physical proof for spirit both through subtle matter and in the broader phenomenon of magnetic rapport (in his unpublished work of 1813, the "Tagebuch über den animalischen Magnetismus").

Technology and the Practice of Spirit

The practice of spirit is not readily separable from the problem of technology. Such a claim should not be surprising in view of our earlier discussion of Fichte's attempt to grapple with the very relation between the spirit and the letter. If one recalls (from Chapter 3), Fichte's correspondence with Schiller showed Fichte's aspiration that the communication of spirit be immediate— all while begrudgingly acknowledging that spirit perhaps indeed required the mediation of the letter. And the very discourse of the letter (that is, writing), it was argued (through reference to Havelock and Ong), must itself be understood as a form of technology.[11] This argument was also developed through reference to Plato's *Phaedrus* and though an analysis of the relation between

---

11. See Chapter 2.

philosophy and rhetoric. It was explained that at its very inception, in Athens, philosophy began with the separation of *techné* from *epistemé*. Such a distinction was established in order to preserve knowledge (*epistemé*) from the sophist's attempt to reduce *logos* in the form of rhetoric to a mere instrument of persuasion.[12]

While Fichte too would attempt to preserve such a distinction, his exchange with Schiller highlights the complication the technology of the letter poses for spirit. In fact, through reference to Heidegger, we already noted that the central problem technology poses for spirit is that modern technology is widely viewed as the summary completion of metaphysics. Fichte's search for a practice of spirit as the physicalization (*Physicirung*) of the metaphysics of imaging in the technique of magnetic psychology leads his own metaphysics into a similar crisis.

Fichte first sees magnetic psychology as a practice that embodies the immediacy of spirit—the immediacy of a spirit to spirit communication. Like his earlier grappling with the issue of the spirit and the letter, however, he ultimately sees it as an unacceptable form of mediation and, at its worst, as an instrumentalization of the *logos*. Still, the relation of technology and metaphysics and the thought of technology as a possible culmination of metaphysics become all the more difficult and complicated in view of the historical relation between idealism and materialism.

Fichte's insistence on the dual series of the real and the ideal, his search for a "Physicirung des Idealismus" (in the "Tagebuch über den animalischen Magnetismus"), and even his articulation of the "I" as sheer activity, is not just a response to the pressures of scientific materialism; I believe these reactions also stand as his articulation of a form of materialism, as a material practice, within his own system of the Wissenschaftslehre.[13] While Fichte does not develop an explicit philosophy of technology, I believe his interest in technology is coincident with his interest in spirit; it stands as a practical, empirical account of the social relation. Fichte's interest in technology then, and perhaps the form of materialism he could be understood to develop implicitly, can best be understood in Marx's assertion that "technology . . . lays bare the process of production of the social relations of . . . life."[14] And it is for this reason that spirit cannot simply ignore the letter, nor metaphysics

12. Châtelet, *Platon*, 60–61. See Chapter 1.

13. See, for instance, Fichte's discussion of the realism and materialism that stands at the heart of the Wissenschaftslehre in the "Second Introduction to the Wissenschaftslehre" (*IWL* 38).

14. Marx, *Capital*, 406 n. 2.

technology. A detailed analysis of a possible completion of Fichte's metaphysics, through the material practice of the technique of magnetic rapport, outlines a technology of the imagination that must begin with its first displacement from the transcendental subject to an external metaphysics of imaging.

The Initial Displacement

The first displacement of the imagination refers to a development in Fichte's work from a transcendental to a metaphysical conception of the imagination. As we saw in earlier chapters, the faculty of the imagination defined a unique creative source and power of the transcendental subject. Fichte's account of the imagination in the last period, however, defines the subject as a mere conduit or screen in the larger process of imaging. The power of imaging (*Einbildungskraft*) is no longer a power belonging to the transcendental subject; rather, it is a force beyond it—in which it plays but a limited role.

The source of this transition is Fichte's reformulation of the self's relation both to itself and the other in terms of feeling or affection. The self-relation I am speaking of, of course, turns on Fichte's reformulation of the problem of reflection in terms of the immediacy of self-feeling, which Fichte transformed in the last period of the Wissenschaftslehre in terms of the seeing of seeing (*das Sehen des Sehens*) or, in short, an immediate reflexibility (*Reflexibilität*) (*FW* 9:543, 10:489).

I begin with the problem of reflection—which Fichte formulates in the last period in terms of a self-seeing eye. While I offered a reading in the previous chapter that attempted to show how in the first period the ego arose anterior to the structure of reflection, the phenomenal ego was nevertheless theorized by means of a structure of reflection. Through a dynamic of reflection akin to the Kantian transcendental unity of apperception the "noumenal" self-feeling ego turned back upon itself and grasped itself as object. Here what was produced as object was the transcendental subject.

While, of course, I have been arguing that the dynamic of self-feeling exhibits an immediate self-awareness prior to a transcendentally conceived self (and consequent ascension to a full-blown self-consciousness—which defines the transcendental subject), this moment of reflection nevertheless represents the dominant perspective from which Fichte develops the Wissenschaftslehre during the first period. In this period, self-conscious reflection stands as a process of self-representation. What changes in the last period with the idea of the self-seeing eye is the very dynamic of mirroring inherent in reflection. Here the act of mirroring not only reflects, but is also the very locus of seeing. Thus, reflection, which was passive, is now simultaneously an active self-seeing. As a consequence, in the *Thatsachen des Bewußtseins* (1813), the problem of reflection is transformed and in many ways resolved, as we will see, in the idea of reflexibility (*FW* 9:543).

In the *Wissenschaftslehre* of 1812, "reflexibility" is also the term Fichte uses to describe that phenomenon beyond mere reflection known as the seeing of seeing (*das Sehen des Sehens*) (*FW* 10:489). Reflexibility is the act of the I as a self-seeing eye. It is *not* a noumenal subject that comes to grasp itself as an object through representation. Rather, like the idea of the self-feeling self, and as a deepening of that project, the self-seeing eye describes the immediacy of the subject-object that is at once indistinguishably producer and product, the seeing and the seen. As such, the reflexibility of the self-seeing eye describes the I as a juncture that has exceeded the framework of the reflective understanding.

The idea of a self-seeing eye fascinated Fichte from the period of the *Wissenschaftslehre Nova Methodo* (1796/99) until his death in 1814.[15] It is in the *Nova Methodo* that Fichte first realizes that "the I of the *Wissenschaftslehre* is not a mirror; it is an eye. . . . The eye . . . is a self-mirroring mirror" (*WLnm* 151). This idea would come more and more to dominate Fichte's thought. Thus, if in the *Wissenschaftslehre* of 1801 Fichte explicitly describes what was formerly a self-positing consciousness in terms of "an activity in which an eye is inserted," it is the deepening of this idea that seems to mark a point of revelation for Fichte.[16] For instance in an unpublished manuscript from the summer of 1812 we are given a glimpse of Fichte's dream-like revelation. He writes: "August 18. Holidays. In a dream a task shown forth quite brightly to me. Self-seeing eye = reflection of a life, of self-manifestation."[17] Now Fichte's talk of self-manifestation, however, cannot be understood as reflection in the traditional philosophical sense. Self-manifestation does not describe a passive mirroring, but an *active* principle of

15. Dieter Henrich claims that Fichte was "constantly fascinated" with this idea from 1801 to 1814. However, I disagree with Henrich on the dating. I believe that Fichte's remarks in the *Wissenschaftslehre Nova Methodo* show that he was concerned with problem of the self-seeing eye at a much earlier date (1796/99). See Henrich, "Fichte's Original Insight."

16. Henrich, "Fichte's Original Insight," 31.

17. Ibid., 33.

appearance. The self-seeing eye, as simultaneously subject and object, achieves a transparency through this reflexibility. Reflexibility then clearly pushes us beyond reflection, conceived as a function of the discursive under-standing. The question, however, is whether, or to what extent, reflexibility and the self-seeing eye can be understood to be a development toward what Hegel terms absolute or speculative reflection.

From Absolute Reflection to *Reflexibilität*

In the *Differenzschrift* Hegel describes how philosophical reflection is trans-formed and resolved by means of absolute reflection. He writes: "To the extent that reflection makes itself into its object, its highest law . . . is its cancellation. . . . Thus, in order to exist it must give to itself the law of self-destruction."[18] While these remarks of Hegel's show that the highest moment of reflection resides in its own annihilation, he will go on to note that this culminating destruction is also the first emergence of absolute reflection. Fichte's idea of the understanding of the understanding and the seeing of seeing then seems to designate a moment in which reflection makes itself into its own object. But the momentum of the cycle forces each last reflection to itself become a new object of reflection. This involution in which reflection becomes a reflection of reflection, and so on, does indeed lead to a self-destruction, a veritable self-annihilation that factually becomes an annihila-tion of the self and its transcendental faculty of representation. As a conse-quence, there is a breakdown: the transcendental subject becomes a mere transparency (*Durchsichtigkeit*) to the absolute.

Absolute or speculative reflection for post-Kantian German idealism broadly identifies the thought of the unity of the subject and object as an identity of identity and difference within the absolute. In view of Fichte's later work, he too seems to achieve the identity of identity and difference, known as the absolute, through the idea of the self-seeing eye. Through the self-destruction of reflection the I/eye becomes transparent to itself. It becomes a transparent image that stands as the site of manifestation for the self-othering appearance of Being. As a conduit of the appearance of Being the transparent I (*das durchsichtige Ich*) functions as a nexus between worlds: it embodies the identity of identity and difference inherent in the ontological

18. Hegel, *The Difference*, 17.

difference between the original productivity of Being and its phenomenal products. This appearance is not representation understood as a product of reflection, but the original appearance and revelation of Being. Here it is significant to note that the terminology of the transcendental subject has for the most part been abandoned: the subject is no longer producer, and the product is no longer merely the object. Rather, the subject is now merely the passive site for the revelation of Being.

While the term "speculative reflection" has come to be associated with the ultimate caricature of philosophy as the whimsy and fantasy of pure thought, *speculatio*, in fact, together with *contemplatio*, form the Latin rendering of the Greek term *theoria*.[19] In order to understand the true place of Fichte's idea of the self-seeing eye within the idealist project of absolute reflection one needs to recognize, at least in Fichte's case, that what one means by the pure thought of *reflectio* is not necessarily opposed to more bodily or practical concerns. Thus what needs to be explained is how, for Fichte, the pure thought (*reines Denken; FW* 10:455) of absolute reflection is not opposed to the reflexibility (*Reflexibilität*) that originated with the affectivity or self-feeling of the body. In other words, is the abstract metaphysics of absolute reflection best understood as grounded and resolved in the *Reflexibilität* of the self-feeling body? Now while the received reading of the later Fichte describes his metaphysics of imaging in terms of a transformation of self-reflection through the notion of the self-seeing eye, my presentation of this first displacement of the imagination brings with it something new. It attempts to make clear how the bodily notion of self-feeling present in the first period continues to be developed in the last period in terms of the language of *Reflexibilität*.

Just how the concept of reflexibility came to be developed as a *reines Denken* is most easily expressed by reviewing the transition from the self-feeling I of the first period of the Wissenschaftslehre to its articulation in the later Wissenschaftslehre as a self-seeing eye. While the term *Reflexibilität* is from the last period of the Wissenschaftslehre, the relation it describes is clearly present in the description of the affective body as a self-feeling self.

The self-feeling self explored in the earlier chapters delineated an affective immediacy that nonetheless implied some sort of phenomenological distance. Consequently, while simple phenomenological distance would describe the I's self-relation as philosophical reflection, it is the unavoidable conflation

19. Gasché, *Tain of the Mirror,* 42.

and interplay of this affective immediacy and phenomenological distance that comes to define reflexibility (as distinct from reflection).[20] In order to bring into relief how the thought of *reflectio* or absolute reflection is achieved through reflexibility (whose paradigm originated within the affective body's self-relation), we need to turn to the work of the French phenomenologist Maurice Merleau-Ponty, and in particular to his "reversibility thesis" as developed in his last work, *The Visible and the Invisible* (1964).[21] Against the perceived ethereal abstraction of the work of Heidegger and Post-Kantian idealism, Merleau-Ponty sets out to reinterpret the Cartesian *cogito,* and by extension the problem of philosophical reflection, through a phenomenology of the subjective body.[22] In other words, Merleau-Ponty confronts the aporia of reflection through a path that, in effect, interprets the idealist resolution in speculative reflection through the nonspeculative terminology of the affective body. This is also what I am suggesting is at work—at least implicitly— throughout much of Fichte's project.

Merleau-Ponty expresses the idea of reversibility most concisely through the idea of the body's own self-experiencing of touching and being touched.[23] When I touch myself, he explains, I am simultaneously touching and being touched. In this dual moment of action and experience we recognize the ambiguity and interrelation of immediacy and phenomenological distance. Like the idealist definition of absolute reflection, there is a unity or simultaneity of identity and difference. Merleau-Ponty explains the dynamic polarity of immediacy and distance as follows: "A sort of dehiscence opens my body in two, and between it the body seen and the body seeing, the body touched and the body touching, there is an overlapping or encroachment."[24] Merleau-Ponty's description of the reversibility thesis—particularly in the language of seeing and being seen—already betrays our interest in his work for Fichte's own conception of reflexibility. In this sense Fichte's account of the self-

---

20. While Gasché contends that "the immediate is itself a reflective act," and thus inherently speculative, I respectfully disagree with this position. I argue that, in the context of my project, through reference to both Fichte and Merleau-Ponty, immediacy cannot be associated with a simple mirroring. See ibid., 74.

21. Merleau-Ponty, *Le visible et l'invisible.* This last fragmentary work was published after Merleau-Ponty's death in 1961. It is also worth noting that, by virtue of the figure of Maine de Biran, there is a rather direct intellectual lineage from Fichte to Merleau-Ponty.

22. The term "subjective body" is a term coined by Merleau-Ponty. It refers to his attempt to think about the transcendental and epistemological subject in terms of embodiment and a phenomenology of activity.

23. Merleau-Ponty, *Le visible et l'invisible,* 165.

24. Ibid.

seeing eye, of the simultaneity of seeing and being seen—like Merleau-Ponty's reversibility thesis—is a description of the affective body. If Fichte resolves the aporia of philosophical reflection in what is called speculative reflection, this absolute reflection is a pure thought (*ein reines Denken*) that nonetheless is inseparable from the being of the subjective body. In other words, what Fichte means by *reines Denken* is not truly distinguishable from his understanding of reflexibility, a reflexibility that originates in the (self-) affectivity of the lived body.[25]

The reflexibility of the self-seeing eye is what makes possible its transparency and passivity before the absolute; and it is this passive disposition which, for Fichte, signals that post-epistemic thought he terms *"reines Denken"* (*FW* 9:432).[26] If the reflexibility of this self-seeing eye is a form of absolute or speculative reflection, then what Fichte means by pure thought cannot reference some ethereal abstraction, but rather our own immediate self-affective self, otherwise called Being, or life.[27]

In the same way that we concluded earlier that self-feeling was the effect of the other, that auto-affection was at heart a hetero-affection, here too in the framework of the last period of the Wissenschaftslehre, like the feeling of feeling, the *Reflexibilität* of the seeing of seeing articulates the subject here as a pure transparency (*ein durchsichtiges Ich*) to the other. Yet what is the nature of this transparent image?

As Miklos Vetö notes, the notion of the image for Fichte—*das Bild*, in the verbal form of *bilden*—conveys "production, affection, and movement. . . . [I]t is not an imagined image [*image imagée*], but an imaging image [*image imageant*]."[28] Such language expresses the *activity* of revelation. However, insofar as the image is *not reproductive*, but actively productive, it is a medium that nevertheless has some immediate relation to the absolute. This is achieved through its transparency.

Thus, the appearance, or image (*das Bild*), for Fichte, acknowledges its own inadequacy and contingency under the form of transparency (*Durchsichtigkeit*). In fact, it can be said to contest its own power of representation.

25. Here, pure thought (*reines Denken*) is achieved in immediacy. In this sense the idea of speculation as merely another form of reflection or mirroring falls apart, insofar as immediacy cannot be truly understood as a mirroring. This is my point of disagreement with Gasché.

26. Wilhelm Wurzer emphasizes the idea of post-epistemic thought in his article "Fichte's Paragonal Visibility."

27. Henry, *Essence of Manifestation*, 77–80. This is Henry's central thesis: the revelation of Being must be understood as affectivity or life.

28. Miklos Vetö, "Les trois images de l'absolu," 39.

As transparent, the image is self-effacing. The transparency is a partial self-annihilation that points beyond itself to the Being beyond appearance. Fichte writes: "Being-there [*Dasein*] must itself grasp, recognize, and image [*bilden*] as pure existence, and must itself—through an absolute Being—posit and image [*setzen und bilden*] . . . it must annihilate itself through its Being by means of another absolute existence [*und muss, durch sein Sein, einem anderen absoluten Dasein gegenüber, sich vernichten*]; which gives it this character of a pure image, of representation [*Vorstellung*] . . . of Being" (*FW* 5:441).

If Fichte seems to use representation and image here interchangeably, it is, I would argue, not because the image is a representation, but the inverse: representation has now come to be understood phenomenologically in terms of the transparent image. Thus, if representation is a function of the self-manifestation of Being, as *Bild*—and "representation" (*Vorstellung*), in the words of Heidegger, "presides over all modes of consciousness"—then the image itself and its process of imaging is already the activity of consciousness (albeit, not the activity of a merely representational consciousness).[29] Fichte explains the intimate relation of representation, consciousness, and image as follows: "The existence of Being is the consciousness or representation of Being" (*FW* 5:441).[30] In other words, to have consciousness at all is a function of representation—representation understood as that transparent image that is the effect of Being's self-othering manifestation. Consciousness then is not representation in the Kantian sense, but an original presentation. The original revelation of Being, as a coming to appearance, is what one means by consciousness: it is a function of the transparency of the image. It is a revelatory, post-epistemic consciousness.

Now it should be clear that the self-other relation of the first period of the Wissenschaftslehre in which the other played a mere ancilliary role—as *Anstoß* and as transcendental condition for the ascension to self-consciousness in the *Grundlage des Naturrechts*—is reversed. The ego, which was earlier defined as act, is not reinterpreted in the last period as a structure whose primary mode is that of passivity. Here we see a first indication of a reciprocity between image and act that is crucial for understanding our own media- and image-saturated environment.

The Other's relation to the I will occur through a revealing process of imaging in which the I stands as a transparent conduit to the radically Other,

29. Henry, *Essence of Manifestation*, 83.
30. Cited in ibid., 80.

the Absolute.[31] Now while our account of the self-other relation in terms of feeling and imaging certainly begins to define the very shift from a transcendental to a more metaphysical account of the imagination, we must continue to establish the groundwork for a more complete explanation in the transition in the dynamic of the imagination through a further understanding of our relation to the other in terms of imaging. In short, we must ask, what is the relation between affect and image? Or, in other words, how does Fichte understand our self-other relation in terms of affection in the development of the imagination as a metaphysics of imaging in the last period of the Wissenschaftslehre?

Image and Affect: A Post-epistemic Consciousness

Affectivity, as original self-feeling, was first seen in the self's process of self-limitation. Just how affectivity is related to this image or phenomenal manifestation should be clear insofar as Fichte describes the power to produce images as a dialectical imagination that arises in the oscillation (*schweben*) between striving and limitation. If one recalls from the analysis of the dual series, this limitation, as self-limitation, is only the idealistic interpretation; its realistic interpretation is said to arise as the effect of an actual other. From this perspective, the very dialectic of the dialectical imagination has a strong social component. In other words, our critical analysis of self-conscious reflection and intersubjectivity, read from the prereflective perspective of auto-affection and hetero-affection, extends to the issue of the imagination. As we will see, then, Fichte's search for a physical proof for the metaphysics of imaging in the social phenomenon of magnetic rapport establishes the imagination as an affective phenomenon that is profoundly social in nature. Once again, following out the logical consequences of the structure of the dual series represented by Fichte's very search for "proof" in rapport, one might suggest that Fichte's quest for a "Physicirung des Idealismus" of his metaphysics of imaging leads him to an affective account of his phenomenological ontology of revelation.

In fact, in his *Die Anweisung zum seligen Leben* (1806), Fichte defines

---

31. Following the convention established by the work of Emmanuel Levinas, I use the term "other" to refer to an empirical individual, and I use the capitalized version "Other" to refer to the radical alterity of the absolute. For the relation of Fichte's thought to the work of Levinas, see my article "Levinas Face to Face with Fichte."

Being, not in terms of an abstract ontology, but rather through a phenomenology of affect. As he explains, "The body is the affect of Being" (*FW* 5:498). The phenomenal manifestation of Being occurs in and through the affective body: Being appears as affect. Affect defines "the feeling of Being as Being" (*FW* 5:498). One's experience of Being as such (and thus of the absolute) can only be experienced through the affective body. Here Fichte's newly developed metaphysics, which articulates a phenomenological ontology of revelation through the affective body, seems to mirror, elaborate, and offer a metaphysical context for his affective description of the social (which, I have argued, grounded his transcendental account through reference to his discourse of subtle matter).

For Fichte's later metaphysics, the language of affect does not exclude, but rather complements, the more traditional language of revelation grounded in the optical metaphors of "seeing" and "imaging." Thus, Fichte can claim that the affective self-feeling self, or the self-seeing eye, is the means by which Being comes to manifestation. In this sense, however, affectivity—as the very source of *Erscheinung*—is to be aligned with the image or the process of imaging. Michel Henry—although in another context—concisely expresses this subtle point helpful for our own understanding of Fichte's later metaphysics. He notes quite simply—"The image is affective."[32] His point, like Fichte's, is that "affectivity is the original essence of revelation."[33] According to Henry, "The affectivity of the image must not be sought in its representative content," but rather "the power which they have of touching and of moving us must be thought of and grasped in itself, as a pure possibility of the a priori order. The affectivity of the image is its internal possibility . . . of Being and its essence."[34] Michel Henry's language of image/not-image describes the same dynamic of phenomenalization as Fichte's vocabulary of Being/Not-Being (*Sein und Nicht-Sein*) (*FW* 10:488). For Fichte, the absolute's process of phenomenalization is understood as follows: the absolute, itself unable to come to appearance, is said "*not* to be," while everything which does come to phenomenal appearance is said "to be." Likewise, for Henry, the image describes all that exists as phenomenal appearance; and what comes to appearance, that is, the very stuff of appearance, is described in yet another register as affectivity. Through reference to Henry, of course,

32. Henry, *Essence of Manifestation*, 487.
33. This is Henry's citation of Fichte. See ibid., 461.
34. Ibid., 488.

I am attempting to make explicit in Fichte's own work the connection between his own vocabulary of Being and Not-Being in his later metaphysics of imaging and his implicit discourse of self-feeling, or affect, that runs—I have been arguing—from his earliest prereflective account of the I as a self-feeling self to his account of the social in terms of subtle matter in the *Grundlage des Naturrechts* and in the "Tagebuch über den animalischen Magnetismus." It is through such a connection that a post-epistemic proof becomes possible.

Affectivity and image are the means by which Fichte is able to articulate a post-epistemic consciousness, a consciousness that is not distinct from the phenomenality of Being. Consequently, Fichte can affirm that "consciousness is absolute existence or the manifestation or the revelation of Being in its only possible form."[35] Consciousness read in terms of the self-seeing eye is an image, an imaging or revelation of Being in its process of self-externalization. Yet while we have suggested that Fichte articulates a post-epistemic cognition beyond the framework of the transcendental subject, we have yet to detail how this achieved.

In Fichte's later work the eclipse of the transcendental subject through the reconception of the "I in terms of the reflexivity of the self-seeing eye results in a self whose sole vocation is to be the site for the revelation and self-manifestation of Being. Fichte affirms in the *Staatslehre* (1813): "Only God is. Outside of him there is only appearance" (*FW* 6:431). If in the first period of the *Wissenschaftslehre* the ego was defined by action, its only possible action is now through imaging. Its power resides in its task of bringing things to appearance. As the self-seeing eye, as producer and product, the I/eye stands as *das Band* or nexus between visible and invisible worlds. Fichte asks, "What was the link of unity [*das Band der Einheit*]?" He answers: "Simply self-seeing [*das Sichsehen*]: I-ness [*die Ich-heit*]. This was thus the point of unity [*Einheitspunkt*] which unifies the highest and absolute opposition [*Gegensatz*]" (*FW* 10:488).

Merleau-Ponty's reversibility thesis, which reinterpreted the power of speculative reflection through a more bodily reading of the phenomenological event by uniting the visible and the invisible through the reversible act of touching and being touched, can again be witnessed earlier in Fichte's designation of the I/eye as the point of "absolute unity of Being and Not-Being [*Sein und Nicht-Sein*]" (*FW* 10:488). Thus, much like Michel Henry's talk

35. Cited in ibid., 76.

of the image and the not-image, Fichte's discussion of Being and Not-Being, as the means by which the invisible is brought to visibility, and Not-Being to phenomenal being, is precisely a description of his later account of the imagination.

Imagination in the Last Period of the *Wissenschaftslehre*

The foregoing critique of reflection is also unavoidably a critique of representation and the representational structure of the transcendental imagination. In the first period of the Wissenschaftslehre the self-positing subject seemed to have an affective origin in the self-feeling self, but it could not account for its own origin outside the mere assumption of its emergence *ex nihilo*. The ego appeared as a consequence of an original act: the act of self-positing. For the early Fichte this original act, which "lies at the basis of all consciousness and alone makes it possible," can therefore "be neither proved nor defined" (*FW* 1:91/*SK* 93). Nevertheless, as Fichte's thought developed—arguably, in part, under the influence of Schelling's critiques—he became more and more interested in the power of the Not-I in the dialectic of the I/ Not-I.[36] Fichte's shift in terminology from the dialectic of the I/Not-I to Being/Not-Being shows an explicit move away from his earlier emphasis on the subject—the self-positing subject—as the creative producer. Thus the imagination, as the productive source, is no longer a faculty situated *within* the transcendental subject, but, in the later work, is aligned with the very source of Being. Still, the self-affective self plays an important role in the imaging power of the imagination. Nevertheless, while we have spoken of the new role of the I/eye as site and transparent image for the manifestation of Being, we have not yet developed in detail this dynamic of imaging as a unique re-articulation of the imagination. This is the task before us.

As early as the *Wissenschaftslehre* of 1794, Fichte located transcendental spontaneity within the imagination, thereby relegating the Kantian conception of the understanding to a "quiescent, inactive power of mind" (*SK* 207).[37] Thus even the early Fichte is witnessed displacing the epistemic function of the understanding to that of a "mere receptacle" (*SK* 207).[38] It is the

---

36. One might argue, however, that this shift was natural since Fichte maintains that what he sought was not the I but merely the unconditioned first principle of knowledge (*FW* 1:91) For Schelling's critique of Fichte, see his essay "On the I as Principle of Philosophy" (*AS* 1:39–134).

37. Cited by Rudolph Makreel in his essay "Fichte Dialectical Imagination."

38. Ibid.

oscillation between dialectical opposition (specifically the opposition between the I and the Not-I) of the synthetic power of the imagination that makes experience as such possible. One should note that with Fichte the imagination is never the servant of knowledge (*Wissen*) as it was for Kant, but rather the source of experience. As we recounted in earlier chapters, the dialectical oscillation (*schweben*) is a centrifugal, centripetal dynamic whereby an outward movement is ultimately reflected back upon itself through the process of striving and limitation. This dialectic—in which oppositions do not truly differentiate—establishes a structure of reflection that makes representational consciousness possible.

Since in Fichte's early work the oscillating imagination was constituted by the opposing elements of self and world, I and Not-I, it was understood as a faculty located within the transcendental subject. In other words, there are not merely indeterminate experiences; but as reflection assures, there are experiences experienced by a particular (experiencing) subject. With the displacement of the transcendental subject in the later theory of the imagination, and the shift to the terminology of Being and Not-Being, the imagination no longer clearly functions as a faculty. The centrality and unity of the subject is called into question.

Yet if the imagination is no longer a faculty of the transcendental subject, how is it to function? The transcendental subject is no longer the producer and creative source, but merely the site of revelation, the conduit of that absolute power beyond it. In Chapter 4 I identified the imagination as part and parcel of the very activity of the self-active and self-constituting character of the I. In the productive imagination the schematization of the dialectical process of striving and limitation arose in the very activity of the I. Here something akin to the Kantian transcendental unity of apperception is achieved through the schematization of the imagination in the activity of the I itself. Now admittedly, from a Kantian perspective, it is not altogether clear whether one could readily elide (without difficulty) the experiential unity provided by the schematism with that provided by the transcendental unity of apperception.[39] Nevertheless, this is precisely which Fichte seems to do.

39. Henry Allison suggests that Kant, at times, may be compelled to elide the unifying function of the schematism with that of the transcendental unity of apperception. He writes, "It would seem, however, that we cannot similarly infer that whenever there is a unity of consciousness there is a representation of an object. Yet this is precisely what Kant appears to be claiming. Indeed, it is what he must claim, if he is to establish a necessary connection between the categories and the unity of consciousness." See Allison, *Kant's Transcendental Idealism*, 146.

It is Fichte's relegation of the understanding to a mere maidservant of the imagination that allows him to develop his conception of a self-active, self-feeling self in terms of the spontaneity of the imagination. As I explained in Chapter 4, in activity, the activity of spatialization, the process of the schematization of the imagination arises in such a way that the schematizing activity of striving and limitation gives unity to that activity we call the I.

It is worth emphasizing that Fichte's description of the I in terms of the dialectical imagination sets it apart from Kant's account in two central respects. First, in contrast to the mere logical subject of Kant's transcendental unity of apperception, Fichte offers a phenomenological account of the ego in terms of sheer activity. Second, in place of what Fichte would view as the artificial stability of the Kantian logical I, his self-active self arises as a quasi-stable image that is the effect or product of the tension and schematization of the dialectical imagination.

In the last period of the Wissenschaftslehre, the self-active self of the first period, which arose as a function of the schematization, is further trans-formed. With the more general shift in emphasis in the last period from the autonomy of the transcendental subject to Being, the very site of thinking, activity, and imaging is altered. Here Fichte directly links what he would formerly describe simply in the language of "act" in terms of "image" or "transparency."[40] Having moved from the discourse of act to image, the self is no longer describable in terms of the self-active motor of the imagination. Rather, as image, the self stands as a *Durch,* as an "in-between," a schematiz-ing mediation between Being and appearance (*Erscheinung*). It is the process of imaging itself then that seeks to mediate "the split between Being and image."[41] The *Einbildungskraft,* as a power (*Kraft*) of imaging, occurs in and through ontological difference, through the tension between Being and its phenomenal manifestation. Through the transparency attained in the self-seeing eye, the Fichtean I/eye is now itself that power of imaging—also known as the schematism—and thereby serves as the *Mittelpunkt* between finite and infinite, sensible and intelligible worlds.

It is from the act of self-seeing that the law of imaging (*das Gesetz des Bildens*) arises (*FW* 9:432). This reflexibility, also conceived as the under-standing of the understanding, effects a reciprocity by means of this transpar-ency which bridges the split between Being and image (*FW* 10:76). And it is

---

40. Dreschler, *Fichtes Lehre vom Bild,* 315.
41. "Die Spaltung in Bild und Sein." Ibid., 326.

out of this bridge—the bridge established by means of the process of imaging—that the schematism arises. Fichte writes, "A schema originates through the split [*Spaltung*] of the understanding itself into image and Being. . . . A schema is the law itself in the Being of the image, or the form of intuition. What the understanding has in each split, is always a law [*ist allemal ein Gesetz*]" (*FW* 10:77). What needs to be emphasized is that this *Spaltung* or self-splitting of the understanding is what makes possible the origin of the schematism and in this sense is *das Gesetz des Bildens.* The *schematism is the law of imaging* that makes possible any appearance whatsoever, and is thus the "uncircumventable" law of spirit.[42] Before we are ready to return explicitly to the question of spirit, however, we must continue to develop and explore its condition of possibility in the schematism.

The I, as *Durch,* makes schematization as such possible. Schematism, in the form of a dialectical image, offers a temporary hiatus or standstill in the dialectical process of separating and relating.[43] Here, the self or I, as *Durch,* is a pure transparency, an image of the image of the absolute. As such, it is to be identified with neither the Kantian transcendental unity of apperception, nor with the activity that stood in for the unity of apperception in the first period of the Wissenschaftslehre. As *Durch,* the self is the process of schematization itself, visible only as a snapshot, or momentary representation in the imagination's continuous process of separating and relating.[44] It is a snapshot of the phenomenal distance and tension between Being's essence and appearance.

For those who might complain that Fichte's account of the I in the last period in terms of the schematism of the imagination no longer seems to allow for an epistemology in terms of adequation, in which a self-identical subject comes to know objects within his or her world, one must recognize that what Fichte now offers us is post-epistemic thinking.[45] With his notion of a pure thinking (*reines Denken*) the transparent subject moves beyond determinate knowing and aligns itself with the absolute in such a way that its primary concern is not epistemological, but ethical.

In fact, it is worth noting that the reflexibility of the I/eye exhibits a synthetic power that bears on both epistemological and ethical considerations. From the perspective of epistemology the I/eye forms the synthetic

42. Cited in Wurzer, "Fichte's Paragonal Visibility."
43. Kinslaw, "The Being of Appearance," 133.
44. Dreschler emphasizes that the term is "Stellvertreter." *Fichtes Lehre vom Bild,* 283.
45. Wurzer emphasizes this point in "Fichte's Paragonal Visibility."

unity of both concept (*Begriff*) and intuition (*Anschauung*), such that *Wissen* is now a function of *Sein*. The epistemic content of seeing has been forever exceeded. While Being and imaging are forever distinct, delimiting the phenomenal distance of space—that very condition of possibility of the imagination insofar as schema equals shape—imaging is supported by Being.[46] Admittedly the schematism, as one's actualizing relation to the absolute, is also intended to be understood as an ethical act. Like the (post-)epistemological synthesis of intuition and concept, the ethical component urges a synthesis of *Leben* (life) and *Vermögen* (possibility) (*FW* 9:548). The ethical imperative is then as follows: one *ought* to make oneself a transparent image and vehicle of the absolute.

Such an imperative bears directly on the social implications of Fichte's transformation of the role of the self in the schematism in the shift from a transcendental to a metaphysical account of the imagination. To understand the significance of such a transformation one must realize that Fichte already conceives all schematization in terms of space, human spatiality, and social practice. Indeed, the word "schema" is etymologically linked to its earlier reference to shape, and by extension, space, insofar as schema is what gives shape to an otherwise indeterminate space.[47] If one recalls from Chapter 4, Fichte would set aside Kant's account of the schematism in terms of an abstract time (and space) in order to develop a schematization through the lived space of human activity. It is in this sense that Fichte can be understood as attempting to overcome what Henri Lefebvre complains of when he writes: "The quasi-logical presupposition of an identity between mental space and (the space of philosophers and epistemologists) and real space creates an abyss between the mental sphere on the one side and the physical and social spheres on the other."[48] Now Fichte's attempt to prevent a schism between the mental and social spheres is clear in both his conception of the schematism in terms of activity in the first period, and his articulation of the world in terms of the spatiality of the image in the last period. As I argued in Chapter 4, in place of the Kantian transcendental unity of apperception, Fichte unified experience through a schematism rooted in the spatiality of activity as such. Experience was unified not in the unity of time, but in the gathering of

46. Ibid.
47. Lefebvre, *Production of Space*.
48. Ibid., 6.

space.[49] The transcendental imagination achieved the power to produce dialectical images through a gathering of the dialectical tension of the spatializing activity of the I.

What changes in the last period with the shift from the transcendental subject is that the schematism, as a process of spatialization and image-making, is no longer an activity of the transcendental subject. Rather, the dialectical tension, formerly between striving and limitation, and now between Being and appearance (*Sein und Bild*), is manifest in the I as such as mere transparency. *The activity of spatialization is now one with the very power of imaging.* In the same way that the activity of spatialization had once unified divergent moments of space, this process is now achieved through the process of imaging as such. Here Fichte confirms a reciprocity between activity and imaging, between the metaphorical languages of the internal and external worlds.

With the structure of the imagination released from the constraints of its facultative determinations, the transcendental is freed from the framework of subjectivity. The source and condition of experience is thus no longer simply cognitive in the Kantian sense, but rather is a function of the phenomenality of Being. It is in this respect, in the "letting be" of Being, that the *affective* vocation of the subject becomes clear. As Michel Henry concludes: "To be a subject means 'to submit,' it means 'to be.' The Being of the subject is Being itself. . . . The essence of subjectivity is affectivity."[50] Here then more than ever the displacement of the subject is tied to our understanding of the self in terms of self-affection—an understanding, one needs to emphasize, that displaces the more cognitive interpretation of the logical subject by means of a radical passivity that readily leads into a second displacement of the imagination through the vocabulary of material technique.

49. I am thinking specifically of Fichte's transformation of the Kantian example of the "line" in which Fichte shows transcendental synthesis to be, not a function of time, but fundamentally the effect of the activity, which is itself the production of space. See Chapter 4.

50. Henry, *Essence of Manifestation*, 476.

8

The Second Displacement

From a Metaphysical to a Technological Imagination

The second displacement of the imagination is marked by a shift from a strict metaphysics of imaging to an account of the imagination in terms of material technique. If the first displacement of the imagination showed a shift in the very site of the productive power of the imagination as a process of externalization, whereby it moved from the transcendental subject (as a faculty) to Being, the second displacement of the imagination continues to affirm the paradigm of an externalized imagination, but further shows how what was once the exclusive domain of the metaphysics of spirit is readily appropriated by human technique and technology. By showing how Fichte's interest in subtle matter and magnetic rapport, as attested to in the "Tagebuch über den animalischen Magnetismus," was a significant philosophical concern as an empirical proof for his account of the metaphysics of imaging in the Wissenschaftslehre, I will demonstrate that Fichte recognized—if perhaps too late—that this technique of rapport was a material technology that threatened to overcome and transform the very metaphysics it was first intended to prove.

Recalling that Fichte had earlier aligned spirit with the imagination by defining it as the power to produce images (*EPW* 193), one should realize that what is at stake in Fichte's attempt to ground his metaphysics of the imagination in the material technique of rapport is the very structure and shape of spirit itself. In other words, *if the empirical proof upon which Fichte grounds his metaphysics comes to take precedence as a technique that determines the very practice of the imagination, then, despite himself, Fichte leads us precisely to what he was trying to evade at all costs: a technologization of spirit.* And although, as I have suggested, Fichte attempts to retreat from these conclusions—and his reversal seems much in line with his earlier attempt to embrace a pure spirit in the face of the technology of the letter (Chapter 2)—two elements nevertheless seem to work against this movement.

First, there is the momentum of the better part of Fichte's own work from the *Grundlage des Naturrechts* forward. The search for an empirical proof in subtle matter in the *Grundlage des Naturrechts* and in magnetic psychology in the "Tagebuch über den animalischen Magnetismus" seems to align itself with the momentum of the weight of history—carried by the itinerary of materialist science and the ever intensifying integration of reproductive technology into everyday life (for instance, as attested to by Benjamin). And second, although Fichte's objection—that material technique produces a *reproductive* imagination that compromises the autonomy of the individual—is well founded, to the extent he initially seeks a proof for both idealism and his metaphysics of the imagination in the material technique of rapport, it is history, from the perspective of our contemporary concern with the role of technology in the constitution of the social medium, that shows this moment to be perhaps the more important one. In other words, Fichte's unique rapprochement between idealism and materialism in the form of the proof of magnetic rapport represents an important instance in the genealogy of the decline of spirit and the concomitant rise of technology in the constitution of community, so that his initial attempt to fortify the claims of idealism through empirical proof, by uniting the communicative power of spirit with material technique—despite its ultimate failings—is a moment perhaps of greater philosophical and historical importance than Fichte's ultimate denial of such a possibility.

What is still more important to realize, however, is that while Fichte may have finally dismissed magnetic rapport—in its particularity—as a possible material proof for his metaphysics, he never gave up his search for such a proof. And it is from this observation of Fichte's continued grappling with the tension between idealism and materialism, and his ultimate suggestion of a possible overcoming of the Wissenschaftslehre, that Fichte's account of the social in view of spirit must be pursued in view of the question of a possible completion of metaphysics by technology.

The Technique of Imagination and the Aesthetic of Rapport

Fichte continued to develop his earlier interest in subtle matter as an empirical proof from the time of the *Grundlage des Naturrechts* through his exploration of the phenomenon of magnetic psychology in the "Tagebuch über den animalischen Magnetismus." Here too, in the "Tagebuch," magnetic psychology was understood as an empirical proof for idealism. It was conceived, in Fichte's words, as a "Physicirung des Idealismus" (TaM 70), as a physicalization or materialization of the claims of idealism, and therefore (he imagined) would stand as a proof against idealism's detractors.

Now although Fichte's later metaphysics of imaging no longer conceived the imagination as a faculty of a transcendental subject, the role of magnetic psychology as an empirical proof does function much like his earlier understanding of the real and ideal series—as a posttranscendental falsifiability.[1] What Fichte's discovery of a "Physicirung des Idealismus" would mean in light of his later metaphysics of the image is that he would find a material, physical analogue, and thus proof, for his previously philosophical account of the imagination.

While Fichte sees a way to ground his metaphysics in the technique of magnetic rapport in order to fortify it against his materialist critics, he has not yet realized the broader danger such a connection poses. Understanding the imagination in terms of revelation creates distinct problems. For instance, as a power of phenomenal manifestation, the imagination's productive power is in danger of being aligned with the technological power of (re)production.

With the imagination set free from its previous facultative vocation within the transcendental subject, one is compelled to ask about the new site (*topos*) of the power of imaging. *When the power of the imagination becomes freed from its vocation as a transcendental faculty, its power of imaging becomes*

---

1. See Chapter 5.

unhinged from the subject as producer. Imaging is thus that which takes place through a passive subject. The passive subject is no longer an active producer, but a receiver that carries out the schematism as a law (*das Gesetz des Bildens*) given unto it. If schematization, as the law of imaging, is understood as a power of the production of space, grounded not within, but beyond the transcendental subject, then the subject—while a passive participant in imaging—is primarily an experiencing subject. It is the recipient of images determined by the law of imaging. But with a passive, transparent subject what if what comes to revelation through it is produced by an external power, a law of imaging *beyond* that dictated by Being?

Now no longer a power of the transcendental subject, the power of imaging is merely what occurs through the subject. As a result, the productive subject now assumes the role of image- or sign-bearer of the absolute. The problem, however, is that the subject is now *no longer merely* the conduit of the absolute. In fact, the metaphysical description of the subject as a pure passivity, which defines its transparency before the absolute, turns out—in concrete terms—to be a description of the subject as a mere receiver or sign-bearer for *any* foreign influence. Consequently, with the technique of rapport it would be possible to reproduce materially what was previously understood to be the work of the imagination.[2] Yet could the power of imaging and the work of spirit then be carried out in the materialist terms of technique and technology? Initially, Fichte did not recognize the danger suggested by such a question. After all, since his interest in explaining magnetic psychology was to prove the truth of idealism, he could not (at first) readily see how his proof would ultimately undermine all he sought to achieve.

Through the language of action and image this double displacement of the imagination is perhaps best understood in terms of its significance for illuminating that unique genealogy of the impact of technology on the social medium as it arises out of the impasse between idealism and materialism. If in the first period of the Wissenschaftslehre it was activity that gave unity to space through movement, such that movement was what made the imagination's schematization of space (as an image) possible, now, in the last period, imaging itself is interpreted as movement. A pure reciprocity has been achieved: all image is movement, and movement image.

2. From the start Mesmerism was understood in large part to be the effect of (and thus a technique of) the imagination. In his "Observations on the Two Reports of the Commissioners Named by the King to Investigate Animal Magnetism" (1784), Charles d'Eslon notes that the commission determined three main causes of Mesmerism: "touch, imitation, and imagination."

Fichte's point is that with the shift of emphasis in his work from knowing to Being the transcendental subject is now no longer the active source of intentionality. The transcendental subject is no longer engaged in an epistemic game that achieves adequation through the consummation of self-identity. The subject no longer engages in willful intentional acts. Rather, the subject, as a transparent structure of imaging, stands as the schema between Being and appearance. In this capacity, as an image of the image of the absolute, its power of imaging could be read as action or movement—as Fichte does—but the movement it displays as image is not its own: it is a mere conduit of absolute Being.

This radical passivity in which the subject is guided (in part) by an exterior will or intentional structure is visible in one's everyday experience of cinema, or as Fichte noted, in magnetic rapport.[3] For instance, in film, while it is, of course, true that it is the spectator who, in one sense, makes narrative continuity possible, in another sense, the very structure of the narrative is determined in advance so that our intentional relation to various actors and actresses is determined at the outset. While one might object that that critical perspective by which one reappropriates one's own intentionality is always possible, such a truism misses the point. The more fundamental point is simply that images carry within them a powerful intentional meaning whose externalized power of intentionality is in many ways equivalent to, or rival to, the intentional structure of the self-active subject. Such an understanding is certainly not new. Political and religious art has for centuries been a testament to this truth. The problem, of course, is that with the rise of technologies of reproductive media such externally generated intentional structures have come to dominate our social relations.

The reciprocity between movement and image also signals a profound confrontation between idealism and materialism for the philosopher Gilles Deleuze: a confrontation that one might understand on the order of the imagination in terms of two rival powers of imaging, between an imagination conceived as a metaphysical ontology and one conceived technologically. In his work on cinema Deleuze writes: "The historical crisis of psychology coincided with the moment at which it was no longer possible to hold a certain position. This position involved placing images in consciousness and move-

---

3. In fact, it was the subject's transparency in magnetic rapport, its transparency to an external will, that first led Fichte to consider it as a possible physical proof of his metaphysical claims. The fact that this external will, however, turned out to be coercive was what ultimately led him to reject it.

ments in space. . . . But how is it possible to pass from one order to another? How is it possible to explain that movements all of a sudden produce an image . . . or that the image produces a movement. . . ? . . . What appeared finally to be dead end was the confrontation of materialism and idealism."[4] The position that could no longer be held was that of the transcendental subject. For Deleuze, cinema's capacity to reproduce images technologically, whose material medium directly influences—or better yet feeds—the representational structure of ideational consciousness, signals a profound historical crisis for psychology. A psychological crisis, however, whose true point of conflict is found in the impasse between materialism and idealism.

The rise of scientific materialism formed the historical horizon that would force idealism and its account of cognitive spontaneity—through reference to the imagination, and thus the soul—to reconsider the entire project of spirit in exclusively materialist terms. That the history of philosophy tells of the simple and unequivocal eclipse of idealism in the face of the materialist paradigm is perhaps no surprise since history is typically written by the victors. In other words, Deleuze's point through reference to cinema, and mine through reference to the influence of magnetic psychology upon Fichte's later metaphysics of the image, is that the crisis of idealism and materialism—brought to us most forcefully through the problem of technology—cannot be resolved through a simple choice between either the spirit or the letter, but rather only through a unique intersection of spirit and matter that exceeds either paradigm's inherent limitations.

Our particular reading of Fichte's later metaphysics will make this evident in two related ways. First, Fichte's own articulation of his metaphysics of the image—in which he establishes an explicit reciprocity between the terms of action and image—seems to anticipate Deleuze's remark about the unique transition between "images in consciousness" and "movements in space" made possible by the technology of cinema.[5] This leads to our second point. If, as is the case, the philosophy of Fichte is *not* a direct product of the aesthetic of cinema, as was Bergson's (according to Deleuze), Fichte's philosophy is nevertheless the product of that same historical horizon whose impasse between idealism and materialism made possible the birth of the technology of cinema. For Fichte, the technology that embodied the unique

---

4. Deleuze, *Cinema I*, 56.
5. Deleuze also writes: "Movement, as physical reality in the external world, and the image, as psychic reality in consciousness, could no longer be opposed." Ibid., xiv.

impasse of materialism and idealism was the technique of magnetic rapport. And its aesthetic, as we will see, is not altogether different from that of cinema.

Rapport, like cinema, was made possible by the rendering passive of the transcendental subject through a certain external *techné* by which the individual's transcendental powers would be constituted and given direction.[6] What this means is that with the technique of rapport and the technology of cinema, the image is already understood as action. The image is the embodiment of the intentional act. The image as an ideational percept—whether supplied by the technology of the camera or the technique of the magnetizer—already supplies both the intentional structure of action (within that image) and the narrative continuity that, like the Kantian transcendental unity of apperception, gives unity to the subject.[7] Yet how do these material forms of imaging relate to the notion of imaging articulated in Fichte's later metaphysics? Such a question can only be explored by answering how Fichte sought to expand and justify his metaphysics against the onslaught of materialist science through his search for a "Physicirung des Idealismus" in magnetic rapport.

The Technique of Rapport as a "Physicirung des Idealismus"

Fichte composed the "Tagebuch über den animalischen Magentismus" in 1813. In this journal, which he never intended for publication, he analyzes the phenomenon of animal magnetism, otherwise known as Mesmerism, and its attendant theory of magnetic psychology. Mesmerism was all the rage throughout Europe at the turn of the eighteenth century.[8] The debate surrounding Mesmerism was especially intense in Berlin around the time Fichte was lecturing there in 1813. Berlin was a hot spot for Mesmerism because, through the figure of Karl Christian Wolfhart (who trained with Mesmer himself), it was the perhaps the central transmission point on German soil for this doctrine.[9]

6. "Transcendental" understood not in the Kantian, but in the Fichtean sense, as a world-constituting activity: here, there is a reciprocity between action and image.

7. Baudry points out that with cinema "the world will be constituted not only by this eye but for it. The mobility of the camera seems to fulfill the most favorable conditions for the manifestation of the 'transcendental subject.'" "Ideological Effects," 292. This point, and how the intentional structure of cinema differs, for instance, from the narrative structure of the novel, was already addressed in Chapter 1.

8. See Darton, *Mesmerism,* and Ellenberger, *Discovery of the Unconscious.*

9. Léon, *Fichte et son temps,* 2:281.

In his work *Fichte et son temps,* Xavier Léon is one of the first and only scholars to acknowledge the philosophical significance of Fichte's enounter with Mesmerism.[10] His approach to the topic is brief, but balanced. On the one hand, he cites Fichte's concern with certain affinities between the Wissenschaftslehre and animal magnetism, while, on the other, he suggests that Fichte's central task was to debunk such theories that so readily appealed to romanticism through his project of critical philosophy. Yet Léon's account of Fichte's encounter with Mesmerism seems largely guided by the belief that although Fichte does find some aspects of magnetic rapport thought-provoking, his central mission is to subject such phenomena to the rigor of science and critical philosophy. Léon's account, however, is not only brief, it is limited in scope. For instance, its analysis does not acknowledge the full extent of Fichte's involvement with the discourse of magnetic rapport, as manifest in the *Grundlage des Naturrechts,* nor does it make clear that Fichte's critical assessment of the phenomenon began with his consideration of it as a possible proof for the claims of the Wissenschaftslehre.

In the "Tagebuch über den animalischen Magnetismus," Fichte's extensive assessment of the phenomenon of animal magnetism shows an in-depth understanding of the subject and detailed mastery of the literature, which could only have come from years of research. This fact, together with Fichte's references to subtle matter in the *Grundlage des Naturrechts,* suggest he was interested in magnetic rapport for many years before he began writing the "Tagebuch." Another indication of his commitment to an investigation of the possible philosophical importance of magnetic rapport is apparent in the narrative at the beginning of the "Tagebuch," in which he recounts his personal encounter with magnetic rapport or Mesmerism (with Wolfhart) that same year.

Through an analysis of the dual series of the real and ideal in the previous chapter and Fichte's broader transformation of Kantian transcendental philosophy, I suggested that Fichte's appeal to the use of subtle matter in the *Grundlage des Naturrechts* was motivated by the very structure of the demands of his own philosophical system. By the date of composition of the "Tagebuch über den animalischen Magnetismus," however, it is clear that Fichte's search for an empirical verification of his transcendental and/or metaphysical

---

10. Günter Schulte is one of the few others to address Fichte's encounter with magnetic psychology. See his article "'Übersinnliche' Erfahrung als transcendentalphilosophisches Problem." Unfortunately, he ultimately does not take seriously its implications for Fichte project.

claims was not mandated by internal philosophical considerations alone. Fichte's investigation into magnetic psychology was also made necessary by the external demands of scientific materialism. In fact, as Xavier Léon emphasizes, Fichte's investigation into magnetic psychology was necessary because, through the voices of men like Karl Solger, magnetic rapport would ultimately be interpreted as a demonstration that threatened the very validity and structure of the transcendental deduction and its claim to deduce experience.[11] In light of this threat posed by magnetic rapport, Fichte's investigations in the "Tagebuch über den animalischen Magnetismus," at least at the outset, can be understood as a unique defense of idealism, a defense which hoped to interpret the threat of magnetic psychology as a proof for idealism.

For many like Solger, if the materialism of subtle matter and magnetic rapport seemed to offer at least as coherent and convincing an explanation as transcendental idealism, where it won out was in its further capacity to demonstrate such claims physically.[12] In response, Fichte too sought physical proof. And he hoped to appropriate the tools of his intellectual enemies for his own use in order to marshal them in support for his own cause. Fichte believed his investigation in the "Tagebuch über den animalischen Magnetismus" would ultimately make it possible for him to sidestep the threat magnetic psychology (as articulated by Mesmer and *Naturphilosophie*) posed to the *Wissenschaftliche* character of transcendental philosophy, by incorporating that threat (of universal fluid) as the very proof of the system of the Wissenschaftslehre. Thus, although the reversal of the priority of thought and Being in the development of the Wissenschaftslehre showed Fichte to be already moving away from more traditional renderings of the transcendental, what he needed and ultimately sought in the phenomenon of magnetic rapport was physical proof for the metaphysics of imaging or the *Bildungslehre* of the last Wissenschaftslehre: proof that "every image is the expression of the law of Being" (TaM 70).[13] The "Tagebuch über den animalischen Magnetismus" is largely concerned not merely with the phenomenon of animal magnetism, but with the search for a material universal, or in Fichte's words, a "Physicirung des Idealismus" (TaM 70) that could ultimately offer physical

---

11. Solger was part of the romantic movement. He explicitly criticized idealism as purely speculative and as being unable to match the level of proof offered by magnetic psychology's account of the world. See Léon, *Fichte et son temps,* 2:281.

12. Ibid.

13. "Alles Bild ist das Ausdruck des Gesetzes des Sein." TaM, 70.

proof for idealism. It is in this sense that Fichte's encounter with Mesmerism is of great importance for Fichte's very project of the Wissenschaftslehre.

In light of criticisms like Solger's, Fichte realized that *if* his account of reality (*Wirklichkeit*) as an image of the image of the absolute was supposed to correspond to the world in which we live, and not merely be an obscure metaphysical deduction, then what would be needed was physical proof that our reality does in fact have an image-like character. As we suggested at the outset, animal magnetism seemed to offer Fichte's later Wissenschaftslehre the physical analogue that would serve as proof for his metaphysics. The question, of course, is how?

Like hypnotism, in the magnetic relation the will of the patient becomes transparent to the will of the magnetist. The individual is controlled by something beyond him- or herself. Fichte describes this magnetic relation in the "Tagebuch über den animalischen Magnetismus" as a "foreign principle" (*fremdes Princip*); one that is, in effect, "the . . . complete annihilation of selfhood" (TaM 3). The idea of the loss of selfhood and a functioning under a foreign principle of course immediately reminds one of the description of the I in the *Bildungslehre* in which the I is said to function as a medium of the absolute. Here, Mesmerism does seem to offer a physical analogue of the idea of the "transparent I" central to Fichte's later metaphysics of imaging.

Despite these similarities, however, Fichte quickly realized that animal magnetism proper cannot be conceived as a "Physicirung des Idealismus." The problem with animal magnetism is precisely that while it seems to signal "the unity of the many in Being" (*Mannigfaltigkeit in der Einheit im Seyn*) through universal attraction, it is in fact merely a lower form of such a phenomenon (TaM 56).

Between Magnetic and Pedagogic Rapport: Rethinking the Intersubjective *Aufforderung*

By contrast with pedagogic rapport, Fichte is able to clarify why magnetic rapport cannot be understood to truly represent the *Physicirung* of the law of the universal. He writes: "The first discoverer has just as little of the true spirit in itself: it is led to him through a completely unknown law and power to which he relates even in itself as student to teacher. Yet there still appears to be a difference; the universal spirit appears to relate itself differently to the discoverer, the teacher to the student. Thus the discoverer himself appears as the last principle of construction: in the unmediated intuition" (TaM 5).[14] The problem with magnetic rapport then is as follows: the self or the I does exhibit a transparency (*Durchsichtigkeit*); yet this transparency reveals not a transparency before absolute Being, but one merely before another human will—that of the magnetist. From this possible false *Physicirung* in magnetic rapport Fichte seems to move toward pedagogy as a possible alternative.[15]

In contrast to magnetic rapport, Fichte wants to claim that the student/ teacher relation is different because the teacher, before whom the student also exhibits transparency, is not "the last principle of construction: in the unmediated intuition," but is himself or herself a messenger of the universal law of absolute Being. This is the core of the problem. Animal magnetism exhibits characteristics of the universal law of the Wissenschaftslehre's metaphysics of imaging, like transparency, but in effect operates "beyond the rule" (*ausser der Regel*) (TaM 58). It operates beyond that rule dictated by Fichte's metaphysics of imaging. From this Fichte concludes: "In what terrible confusion Mesmer and Wolfhart are in is now clear" (TaM 58).[16]

It is important to understand how this search for a "Physicirung des Idealismus" through magnetic and pedagogic rapport represents a continuation of our earlier interpretation of the *Aufforderung* in terms of subtle matter in the *Grundlage des Naturrechts* (Chapter 5). Fichte's turn to pedagogic rapport, and rapport more generally, must itself be understood as part of his attempt to explain the *Aufforderung* in physical, material terms. A hint to such a relation is given in the *Grundlage des Naturrechts*. Fichte writes: "The *Aufforderung* to free self-activity [*Selbstätigkeit*] is that which one names education [*Erziehung*]" (FW 3:39–40). In pedagogic rapport the student also becomes transparent to an external will. Yet the student, Fichte believes, is not transparent to the will of the teacher per se, but only insofar as the teacher stands in for the absolute. And it is this communicative path that stands as the new model of the *Aufforderung*. Here, one is called to self-

---

14. "Der erste Erfinder hat eben sowenig den richtenden Geist in sich: er wird geleitet durch ein ihm ganz unbekanntes Gesetz, und Kraft, zu welchen er ohnegefahr eben so sich verhält, wie Zuhörer zum Lehrer. Es scheint aber doch noch Ein Unterschied zu Seyn; das allegemeine gestige scheint sich anders zum Erfinder zu verhalten, als der Lehrer zum Schuler. So: der Erfinder erscheint sich selbst als das letzte Princip der Konstruktion: in der unmittelbaren *Anschauung*." TaM 5.

15. While Fichte will test magnetism as a proof for idealism against the seeming paragon of pedagogic rapport, Fichte never seems to consider pedagogic rapport itself as a "Physicirung de Idealismus," and thus as an answer to Solger and the materialists.

16. "In welcher schrekl[ichen] verw[irrun]g Mesmer und Wolfhart sind, ist nun klar." TaM 58.

conscious self-activity through the process of education (*Erziehung*) in which one learns for oneself how to becomes transparent to the absolute.

Pedagogy is central to Fichte's very conception of the Wissenschaftslehre. In fact, as Julius Dreschler confirms, "In Fichte's work the pedagogical problem is irresolvably mixed [*unlösbarverschlungen*] with the philosophical problem."[17] A clear instance of this is the continuous reference to Fichte's later metaphysics of imaging as a *Bildungslehre*. This term, *Bildungslehre*, carries with it both the *metaphysical* notion of the transparent image and the *ethical* demand for moral formation (as *Bildung*). After all, one ought to become transparent to the absolute, and one achieves such metaphysical heights through the concrete moral and educational practice of self-renunciation.

The philosophical problem is a pedagogical problem insofar as for each the issue turns on the self's relation to the world. What defines this relation is the distance to be overcome as a work of spirit. The philosophical account articulates the self's relation to the world as a medium whose transparency offers a communicative immediacy that defines spirit. The pedagogical account bridges the distance between self and world as an *Erziehung* or *Afforderung* that calls the other to a self-recognition of one's immediacy to the world as a mere conduit or transparent image of the absolute. (Although, as I will argue, the very problem with these models, insofar as they remain intersubjective, is that the achievement of spirit is infinitely deferred, or always in process.)

Why Fichte refused to consider pedagogical rapport as a "Physicirung des Idealismus" initially seems unclear. Fichte uses pedagogy to highlight the failure of magnetic rapport. My best explanation for this decision comes down to each respective form of rapport's position vis-à-vis the competing claims of intersubjectivity and spirit. Thus, I can only suggest that Fichte based his decision on the fact that pedagogy or *Erziehung*, as a form of inter-subjectivity (*FW* 3:36), stands as an "ought," as a process yet to be achieved; whereas, by contrast, magnetic rapport had offered a pure transparency, an immediacy that stood as a complete material analogue and proof of the ideal of spirit. While *Erziehung* does exhibit some form of transparency, it is nei-ther immediate nor complete. It is equally a process of intersubjective forma-tion. And, as I have reiterated, Theunissen, Habermas, and others have suggested, it is intersubjectivity's very structure of mediation that stands as a roadblock to spirit.

17. Dreschler, *Fichtes Lehre vom Bild,* 26.

A Technology of Spirit? Proof, Material Technique, and the Completion of Metaphysics

Fichte's pursuit of a "Physicirung des Idealismus" led him to the technique of magnetic rapport. While pedagogy stood as an important model for Fichte, the dynamic of *Erziehung* is fundamentally intersubjective and thus cannot serve as the material instantiation of the absolute communicative transparency and immediacy that is the hallmark of spirit. It is not surprising that Fichte at first saw the technique of magnetic rapport as the ultimate material proof of spirit. It embodied the very ideal of communicative technology: *total transparency*.[18] It was literally a transparent medium, a medium whose invisibility made the unity of spirit possible in material terms. And it is in this sense that metaphysics could be understood to achieve completion in technology. In fact, one might say the dream of communicative technology is the completion of metaphysics. Fichte, of course, had hoped that the technique of rapport would stand as a "Physicirung des Idealismus" and thus would, in effect, complete the dream of metaphysics.[19] His recognition, however, that the ends of the technique of magnetic rapport did not prove or complete his metaphysical claims left him in a rather difficult position.

In an earlier chapter I suggested that Fichte reformulated Kant's transcendental philosophy in such a way that it was dependent upon a posttranscendental move. Although, as a construction of thought, the validity of the Kantian transcendental a priori is either affirmed or denied at the level of the transcendental deduction, Fichte's own hypothesis is tied to a posttranscendental move. Fichte's dependence upon a posttranscendental proof was so significant that he claimed that "if the results of such philosophy do not agree with experience, then the philosophy given is surely false" (*IWL* 32). In view of this position, Fichte's inability to establish a posttranscendental proof of his metaphysics of imaging or spirit leaves his metaphysical claims in a vulnerable position. My point is that it was not the rising tides of materialism alone that demanded a physical proof of Fichte's metaphysics of spirit, but the rigor of his own methodology. If magnetic rapport fails to stand as a "Physicirung des Idealismus"—and his metaphysics stands without a post-

---

18. Don Ihde asserts that the very ideal of technology is total transparency. See *Existential Technics*, 51.

19. And once again, one thinks of Fichte's urging Schiller to somehow embrace the communicative transparency of the spirit of his work through the technology of its letter.

transcendental proof, then we must conclude, according to Fichte's own criterion, that "the philosophy given is surely false" (*IWL* 32).

Such a pronouncement, while seemingly severe, needs to be understood both in the context of Fichte's greater project of the Wissenschaftslehre and in the context of post-Fichtean philosophy. Now although Fichte warns that one should not base philosophy upon experience (*IWL* 32), one imagines that the development of Fichte's philosophy was at least driven by a process of falsifiability.[20] It is nevertheless worth emphasizing that the Wissenschaftslehre was an active, ever evolving, lifelong project that went through at least sixteen major drafts. My point is that Fichte's failure to provide proof for spirit, and his metaphysics of imaging in 1813, did not necessarily spell the end of idealism for Fichte, or even the end of philosophy as Fichte conceived it. For instance, although Fichte's death several months after the completion of the "Tagebuch über den animalischen Magnetismus," prevented him from presenting any new version of his philosophy, in the "Tagebuch über den animalischen Magnetismus," he does hint at the possibility of a "true overcoming of the Wissenschaftslehre" (TaM 87).[21] Although Fichte is not clear in what way the Wissenschaftslehre could be overcome, from the context of his discussion of the import of magnetic rapport for the Wissenschaftslehre, it is likely that he was envisioning a rapprochement of sorts between the competing claims of idealism and materialism—perhaps through a new understanding of a "proof" of idealism.

From a philosophical perspective, one might be tempted to ask both whether an empirical proof is really necessary for transcendental idealism, and whether such a proof would make any sense. For Fichte, an empirical proof was necessary for two reasons. First, philosophically, he believed an empirical proof could be used to falsify inadequate philosophical constructions. In short, the proof kept philosophers honest. And I have argued that the very notion of "proof" for transcendental philosophy was essential to

---

20. To reiterate, Fichte seems caught between two possible alternative types of posttranscendental moves. The lesser claim of falsification only requires that empirical experience does not falsify his transcendental claims, whereas verification, the stronger demand, would require a positive proof for transcendental claims. He seems to want magnetic psychology to stand as this positive proof, as a physical embodiment of his philosophy. At the very least, however, his reformulation of the transcendental as one which requires a posttranscendental move would require—at the very least—the process of an attempted empirical falsification of any transcendental claim. And his philosophy does not seem to fair well, in view of magnetic rapport, under either of these possible interpretations of its own systemic requirements for a posttranscendental move.

21. "Es ist hier eine wahre Erhebung des WL." TaM 87.

Fichte's reformulation of Kantian transcendental philosophy. From his earliest notion of the dual series of the real and the ideal, to subtle matter in the *Grundlage des Naturrechts,* to his last searches for a "Physicirung des Idealismus," I have shown a continuity in Fichte's conception of philosophy as an attempt to keep philosophy, at least in part, grounded in this world. Fichte's second concern with a philosophical proof was historical in nature. Insofar as philosophy is always in part a product of its own time, it is important to recognize that the demand that Fichte supply a proof for his metaphysics became all the more intense with the rise of scientific materialism. In the end, Fichte's inability to supply *any* empirical proof for idealism, of course, had both historical and philosophical consequences. Historically, idealism was eclipsed by the rise of scientific materialism. Philosophically, Fichte was unable to meet the demands of his own system. While, as we suggested, he may have established some obscure proof in what he referred to as "the overcoming of the Wissenschaftslehre," at minimum his inability to provide empirical proof—*any proof*—would demand a radical rethinking of the very criterion he first established.

We now return to the final metaphilosophical question of whether Fichte's own criterion of an empirical proof for (transcendental or metaphysical) idealism makes any sense. Again, Fichte's intent was to transform Kantian transcendental philosophy from what he perceived to be an overly abstract system to a vital, experimental activity. Transcendental philosophy must not be divorced from the world. And while Fichte shunned any philosophy merely derived from experience, he did see experience as crucial in confirming the validity of transcendental claims. I concede that empirical proofs for transcendental claims are difficult to establish; yet following Fichte, it does seem that empirical experiences work best as positive proofs of falsifiability. As I noted in a footnote in an earlier chapter, Fichte's desire for an empirical proof is made even more problematic because although his method allows for empirical falsifiability, a negative proof, he secretly desires and seeks a positive empirical proof, a verification or confirmation, for his transcendental and metaphysical claims.

An example of an empirical proof for a transcendental claim is difficult to come by. Therefore, let me offer the following hypothetical example. Imagine a scenario in which Fichte was able to combine the best of pedagogic rapport with the best elements of magnetic rapport. If, for example, the pedagogic model was not fundamentally intersubjective, but offered immediate, total transparency, then a physical proof for spirit and the metaphysics of imaging would have been provided (at least for Fichte). For those still skeptical of Fichte's use of empirical proofs, particularly in his later metaphysics, let me explain it this way. Fichte's demand for proof guarded him against the very excesses of metaphysics that Kant had warned about in the First Critique. And one might add that post-Kantian idealism's transformation of Kantian philosophy would signal a return to metaphysics; one that could indeed use a confirmation or proof of its speculative claims.

While Fichte did not want to directly derive philosophy from experience, he does seem to be in part sympathetic with the materialist critique of metaphysics as mere unverifiable speculation. Take his continuous reformulation of the Wissenschaftslehre. Now from the perspective of Fichte's metaphysics of imaging, in which the subject stands as a transparent I, the description of the transcendental imagination in the first period, as a faculty, seems wholly inadequate. How then should one view his earlier transcendental claims vis-à-vis his later metaphysics of imaging?

I suggest that the very development of the Wissenschaftslehre can be understood on this model of falsification. Now while we noted earlier that the dialectical character of the transcendental imagination was initially bound with the self-reflective structure of intersubjective recognition, and intersubjectivity itself was manifested as *Erziehung*, I argue that the tension between the competing models of intersubjectivity and spirit present in the dynamic of *Erziehung* ultimately led Fichte to a metaphysics of imaging in such a way that the conflicting social model in *Erziehung* led to a falsification of the strictly transcendental claims of intersubjectivity. And, of course, the *Grundlage des Naturrechts*'s search for the ground of intersubjectivity in the empirical proof of subtle matter could also be understood as a moment of falsification through empirical proof.

Conclusion: Technology and the Dream of Metaphysics

Now it seems to me the power and value of Fichte's use of empirical proof in his philosophy lies in the fact that it required his conception of the imagination and social interaction to take material and historical conditions into account. As we saw in Chapter 1, through reference to Walter Benjamin, I suggested that the technology of reproductive media threatened the very structure of intersubjective recognition and thus the predominance of intersubjectivity as a social theory. Likewise, I would argue that Fichte's demand for philosophical proof assured that his account of the imagination, and by extension his account of the social, remained grounded in the prevailing historical and material conditions.

Without proof Fichte's metaphysics could not achieve completion: it could not be proven, or at least be assured that it would be spared falsification. The proof for his metaphysics of imaging (or spirit) through the technique of magnetic rapport was to stand as a second displacement of the imagination. As we noted, it represented the materialization of an already externalized metaphysical account of the imagination. While one might argue that Fichte ultimately rejected magnetic rapport as "proof," his continued desire to discover a viable "Physicirung des Idealismus," I believe, signals a trajectory in his work that necessarily leads to what I have been referring to as a materialization of the imagination or a second displacement of the imagination.

Like communicative technology's aspiration for total transparency (in the material world), Fichte too sought to achieve the ideal of spirit in material terms through the transparency embodied in magnetic rapport. Again, it should be emphasized that despite the failings of the particular proof of Mesmerism, Fichte's very project of a "Physicirung des Idealismus" can be understood as one that aspires to complete the dream of metaphysics—total (communicative) transparency—through a material technique and technology; indeed, his philosophical system requires it. Such a claim, however, holds only to the extent that technology can be understood to literally "embody" spirit. As we saw in the instance of magnetic rapport, it often does not.

Fichte's own understanding of the role material technology plays in the metaphysics of imaging, I believe, is most clearly expressed in the work of Heidegger. Such an assertion should not be surprising since scholars of the later Fichte have long noted the similarities between Fichte's own metaphysics of imaging and Heidegger's phenomenological ontology.[22]

In "The Age of the World Picture," Heidegger writes, "What is dangerous is not technology. . . . The *essence of technology* . . . is the danger. . . . Thus where enframing [*Gestell*] reigns, there is danger in the highest sense."[23] What Heidegger suggests is that the essence of technology, as modern technology, establishes a representational structure or enframing that limits our experi-

22. See, for instance, the work of Dreschler, Janke, Wurzer, and Henry.
23. Heidegger, *The Question Concerning Technology*, 28; my italics.

ence, or better, prevents our experience of the free self-appearance of Being. And it is in this sense that technology is at odds with what Fichte would refer to as spirit's metaphysics of imaging. What is most significant for our analysis of Fichte, however, is Heidegger's recognition that "technology is not demonic," and should not be understood as opposed to an ontology of revelation per se. Thus, where Heidegger speaks of technology as a "saving power," what he is suggesting is that technology can be understood as a supplement or completion of metaphysics in a positive sense. Insofar as technology need not block the manifestation of Being or spirit, but may indeed stand as a "constellation in which revealing and concealing, in which the coming to presence of truth, comes to pass," it can be understood to have fulfilled the very work of spirit. And Fichte's search for a "Physicirung des Idealismus" had much the same aspiration.

Bibliography

Works by Fichte

Fichte: Early Philosophical Writings. Trans. and ed. Daniel Breazeale. Ithaca: Cornell University Press, 1988.

Fichte: Foundations of Transcendental Philosophy (Wissenschaftslehre) Nova Methodo (1796/99). Trans. and ed. Daniel Breazeale. Ithaca: Cornell University Press, 1992.

Fichtes Werke. 11 vols. Ed. I. H. Fichte. 1834–35. Reprint. Berlin: Walter de Gruyter, 1971.

Gesamtausgabe der Bayerishen Akademie der Wissensschaften. Ed. Reinhardt Lauth, Hans Jacob, and Hans Gliwitsky. Stuttgart-Bad Cannstatt: Friedrich Frommann, 1964–.

J. G. Fichte: Introductions to the Wissenschaftslehre and Other Writings. Trans. and ed. Daniel Breazeale. Indianapolis: Hackett, 1994.

"On the Spirit and the Letter of Philosophy, in a Series of Letters." Trans. Elizabeth Rubenstein. In *German Aesthetic and Literary Criticism: Kant, Fichte, Schelling, Schopenhauer, Hegel,* ed. David Simpson, 74–93. Cambridge: Cambridge University Press, 1984.

The Science of Knowledge. Trans. and ed. Peter Heath and John Lachs. Cambridge: Cambridge University Press, 1982.

Other Works

Adorno, T. W. *Aesthetic Theory.* Trans. C. Lenhardt. London: Routledge and Kegan Paul, 1984.

———. *Against Epistemology: A Metacritique. Studies in Husserl and the Phenomenological Antinomies.* Trans. Willis Domingo. Oxford: Basil Blackwell, 1982.

Adorno, T. W., and Max Horkheimer. *Dialectic of Enlightenment.* Trans. John Cumming. New York: Seabury Press, 1972.

Allison, Henry E. *Kant's Transcendental Idealism: An Interpretation and Defense.* New Haven: Yale University Press, 1983.

Baudry, Jean-Louis. "Ideological Effects of the Basic Cinematographic Apparatus." In

Narrative, Apparatus, Ideology: A Film Theory Reader, ed. Philip Rosen. New York: Columbia University Press, 1986.

Baumann, Peter. *Fichtes Ursprungliches System*. Stuttgart: Frommann-Holzboog, 1972.

Benjamin, Walter. *Gesammelte Schriften*. Frankfurt: Suhrkamp, 1980.

———. *Illuminations: Essays and Reflections*. Ed. Hannah Arendt. Trans. Harry Zorn. New York: Schocken Books, 1968.

Berdyaev, Nicolas. *Spirit and Reality*. Trans. George Reavey. London: Geoffrey Bles, 1939.

Bergson, Henri. *Matter and Memory*. Trans. N. M. Paul and W. S. Palmer. New York: Zone Books, 1988.

Borch-Jacobsen, Mikkel. *The Emotional Tie: Psychoanalysis, Mimesis, and Affect*. Trans. and ed. Douglas Brick et al. Stanford: Stanford University Press, 1992.

Breazeale, Daniel. "Check or Checkmate? On the Finitude of the Fichtean Self." In *The Modern Subject: Conceptions of the Self in Classical German Philosophy*, ed. Karl Ameriks and Dieter Sturma. Albany: SUNY Press, 1995.

Cantor, G. N., and Michael J. S. Hodge, eds. *Conceptions of Ether (1749–1900)*. Cambridge: Cambridge University Press, 1981.

Châtelet, François. *Platon*. Paris: Gallimard, 1965.

Crawford, Donald W. *Kant's Aesthetic Theory*. Madison: University of Wisconsin Press, 1974.

Darton, Robert. *Mesmerism and the End of the Enlightenment in France*. New York: Schocken Books, 1970.

Deleuze, Gilles. *Cinema I: The Movement-Image*. Trans. Hugh Tomlinson and Barbara Habberjam. Minneapolis: University of Minnesota Press, 1991.

Dreschler, Julius. *Fichtes Lehre vom Bild*. Stuttgart: W. Kohlhammer Verlag, 1955.

Druet, Pierre-Philippe. "L' 'Anstoss' fichtéen: Essai d'élucidation d'une métaphore." *Revue Philosophique de Louvain* 70 (1972).

Düsing, Edith. *Intersubjektivität und Selbstbewußtsein*. Köln: Dinter Verlag, 1986.

Ellenberger, Henri. *The Discovery of the Unconscious: The History and Evolution of Dynamic Psychiatry*. New York: Basic Books, 1970.

d'Eslon, Charles. "Observations on the Two Reports of the Commissioners Named by the King to Investigate Animal Magnetism." 1784. In *The Nature of Hypnosis: Selected Basic Writings*, ed. Ronald E. Shor and Martin T. Orne. New York: Holt, Rinehart and Winston, 1965.

Ferry, Jean-Luc. *Political Philosophy 2: The System of Philosophies of History*. Trans. Franklin Philip. Chicago: University of Chicago Press, 1992.

Finger, Otto. *Von der Materialität der Seele*. Berlin: Akademie Verlag, 1961.

Freud, Sigmund. *Massenpsychologie und Ichanalyse*. In *Gesammelte Werke*, vol. 13. London: Imago, 1947.

Gasché, Rudolph. "Some Reflections on the Notion of Hypotyposis in Kant." *Argumentation* 4 (1990): 85–100.

———. *The Tain of the Mirror*. Cambridge: Harvard University Press, 1987.

Gross, Larry, John Stuart Katz, and Jay Ruby, eds. *Image Ethics: The Moral Rights of Subjects in Photographs, Film, and Television*. Oxford: Oxford University Press, 1988.

Guyer, Paul. *Kant and the Claims of Taste*. Cambridge: Harvard University Press, 1979.

Habermas, Jürgen. "Arbeit und Interaktion." In *Frühe politische System*, ed. G. Göhler. Frankfurt: Ullstein, 1974.

Havelock, Eric. *Preface to Plato.* Cambridge: Harvard University Press, 1962.

Hegel, G. W. F. *The Difference Between Fichte's and Schelling's Systems of Philosophy.* Trans. H. S. Harris and Walter Cerf. Albany: SUNY Press, 1977.

———. *Phenomenology of Spirit.* Trans. A. V. Miller. New York: Oxford University Press, 1977.

Heidegger, Martin. *An Introduction to Metaphysics.* Trans. Ralph Mannheim. New Haven: Yale University Press, 1959.

———. *The Question Concerning Technology and Other Essays.* Trans. Willam Lovitt. New York: Harper & Row, 1977.

Helfer, Martha B. *The Retreat of Representation.* Albany: SUNY Press, 1996.

Henrich, Dieter. "Fichte's Original Insight." Trans. David Lachterman. In *Contemporary German Philosophy I,* 15–52. University Park: Pennsylvania State University Press, 1982.

Henry, Michel. "Does the Concept of the Soul Mean Anything?" *Philosophy Today* 13, nos. 2–4 (1969): 94–114.

———. *Essence of Manifestation.* Trans. Girard Etzkorn. The Hague: Martinus Nijhoff, 1973.

———. *The Genealogy of Psychoanalysis.* Trans. Douglas Brick. Stanford: Stanford University Press, 1993.

Hunter, C. K. *Der Interpersonalitätsbeweis in Fichtes früher angewandter praktisher Philosophie.* Meisenheim am Glan: Verlag Anton Hain, 1973.

Husserl, Edmund. *Cartesian Meditations.* Trans. Dorian Cairns. The Hague: Martinus Nijhoff, 1960.

Huyssen, Andreas. *After the Great Divide: Modernism, Mass Culture, Postmodernism.* Bloomington: Indiana University Press, 1986.

Ihde, Don. *Existential Technics.* Albany: SUNY Press, 1983.

Janke, Wolfgang. *Fichte: Sein und Reflexion.* Berlin: Walter de Gruyter, 1970.

Kant, Immanuel. *Critique of Judgment.* Trans. Werner Pluhar. Indianapolis: Hackett, 1987.

———. *Critique of Pure Reason.* Trans. Norman Kemp Smith. New York: St. Martin's Press, 1965.

———. *Gesammelte Schriften.* Ed. Preussischen Akademie der Wissenschaften zu Berlin. 29 vols. Berlin: Walter de Gruyter, 1902–83.

Kearney, Richard. *The Wake of the Imagination.* London: Blackwell Publishers, 1988.

Kinslaw, C. Jeffrey. "The Being of Appearance: Absolute, Image, and the Trinitarian Structure of the 1813 Wissenschaftslehre." In *New Perspectives on Fichte,* ed. Daniel Breazeale and Tom Rockmore. Atlantic Highlands, N.J.: Humanities Press, 1996.

Kittler, Friedrich A. *Gramophone, Film, Typewriter.* Trans. Geoffrey Winthrop-Young and Michael Wutz. Stanford: Stanford University Press, 1999.

Lacoue-Labarthe, Philippe. *Typography: Mimesis, Philosophy, Politics.* Ed. Christopher Fynsk. Cambridge: Harvard University Press, 1989.

Lacoue-Labarthe, Philippe, and Jean-Luc Nancy. *The Literary Absolute.* Trans. Philip Barnard and Cheryl Lester. Albany: SUNY Press, 1988.

Lauth, Reinhardt. "Le problème de l'interpersonalité chez J. G. Fichte." *Archive de Philosophie* 26 (1962): 324–44.

Lefebvre, Henri. *The Production of Space.* Trans. Donald Nicolson-Smith. Cambridge: Blackwell Publishers, 1991.

Léon, Xavier. *Fichte et son temps.* Vol. 2. Paris: Libraire Armand Colin, 1927.

Lessing, Gotthold Ephraim. *Laocoön: An Essay on the Limits of Painting and Poetry.* Trans. Edward Allen McCormick. Baltimore: Johns Hopkins University Press, 1984.

Levinas, Emmanuel. *Totality and Infinity.* Trans. Alphonso Lingis. Pittsburgh: Duquesne University Press, 1969.

Makreel, Rudolph A. "Fichte's Dialectical Imagination." In *Fichte: Historical Contexts/ Contemporary Controversies,* 7–16. Atlantic Highlands, N.J.: Humanities Press, 1994.

Martin, Wayne. *Idealism and Objectivity: Understanding Fichte's Jena Project.* Stanford: Stanford University Press, 1997.

Marx, Karl. *Capital: A Critique of Political Economy.* Trans. Samuel Moore and Edward Aveling. New York: Random House, 1906.

Merleau-Ponty, Maurice. *Le visible et l'invisible.* Ed. Claude Lefort. Paris: Gallimard, 1964.

Neuhouser, Frederick. *Fichte's Theory of Subjectivity.* Cambridge: Cambridge University Press, 1990.

Olson, Alan. *Hegel and the Spirit: Philosophy as Pneumatology.* Princeton: Princeton University Press, 1992.

Ong, Walter. *Orality and Literacy: The Technologization of the Word.* London: Routledge, 1982.

Patterson, Richard. *Image and Reality in Plato's Metaphysics.* Indianapolis: Hackett, 1985.

Philonenko, Alexis. *La liberté humaine dans le philosophie de Fichte.* Paris: Vrin, 1966.

Pippen, Robert. *Hegel's Idealism: The Satisfactions of Self-Consciousness.* Cambridge: Cambridge University Press, 1989.

Plato. *The Collected Dialogues of Plato.* Ed. Edith Hamilton and Huntington Cairns. Princeton: Princeton University Press, 1961.

———. *The Republic.* In *The Collected Dialogues of Plato,* ed. Edith Hamilton and Huntington Cairns. Princeton: Princeton University Press, 1961.

Ricoeur, Paul. *Time and Narrative.* Vol. 3. Trans. Kathleen Blamey and David Pellauer. Chicago: University of Chicago Press, 1988.

Rockmore, Tom. *Fichte, Marx, and the German Philosophical Tradition.* Carbondale: Southern Illinois University Press, 1980.

Rutsky, R. L. *High Technē: Art and Technology from the Machine Aesthetic to the Posthuman.* Minneapolis: University of Minnesota Press, 1999.

Sallis, John. *Spacings—Of Reason and Imagination in the Texts of Kant, Fichte, Hegel.* Chicago: University of Chicago Press, 1987.

Schelling, F. W. J. *Ausgewählte Schriften.* 6 vols. Edited by Manfred Frank. Frankfurt: Suhrkamp Verlag, 1985.

———. *Sämmtliche Werke.* 14 vols. Ed. K. F. A. Schelling. Stuttgart: Cotta, 1856–61.

———. *The System of Transcendental Idealism (1800).* Trans. Peter Heath. Charlottesville: University Press of Virginia, 1978.

Schulte, Günter. "'Übersinnliche' Erfahrung als transcendentalphilosophisches Problem. Zu Fichtes 'Tagebuch über den animalischen Magenetismus von 1813.'" In *Der transcendentale Gedanke: Die gegenwärtige Darstellung der Philosophie Fichtes,* ed. Klaus Hammacher. Hamburg: Meiner Verlag, 1981.

Scribner, F. Scott. "The Aesthetics of Influence: Fichte's *Grundlage des Naturrechts* in View of Kant's Third Critique." In *Rights, Bodies, and Recognition,* ed. D. Breazeale and T. Rockmore, 161–77. Aldershot, U.K.: Ashgate, 2005.

————. "Falsification: On the Role of the Empirical in J. G. Fichte's Transcendental Method." In *Fichte, German Idealism and Early Romanticism (Fichte-Studien Supplementa)*, ed. Daniel Breazeale and Tom Rockmore. Amsterdam: Rodopi, 2009.

————. "Levinas Face to Face with Fichte." *Southwest Philosophy Review* 16, no. 1 (2000).

————. "Die 'Physicirung des Idealismus' im 'Tagebuch über den animalischen Magnetismus': Die letzte Wissenschaftslehre oder Das Ende des Idealismus?" In *Fichte-Studien (Bd. 17–18): Die Spätphilosophie J. G. Fichtes*, ed. Wolfgang H. Schrader, 319–328. Amsterdam: Rodopi, 2000.

————. "Reduction or Revelation? Fichte and the Question of Phenomenology." In *Fichte and Phenomenology (Fichte-Studien Supplementa)*, ed. Daniel Breazeale and Tom Rockmore. Amersterdam: Rodopi, 2009.

————. "The 'Subtle Matter' of Intersubjectivity in the *Grundlage des Naturrechts*." In *New Essays on Fichte's Later Jena Wissenschaftslehre*, ed. Daniel Breazeale and Tom Rockmore, 65–79. Evanston: Northwestern University Press, 2002.

Seyhan, Azade. *Representation and Its Discontents: The Critical Legacy of German Romanticism*. Berkeley and Los Angeles: University of California Press, 1992.

Siep, Ludwig. *Anerkennung als Prinzip der Praktischen Philosophie*. Freiburg: Alber Verlag, 1979.

Theunissen, Michael. "Die verdrängte Intersubjektivität in Hegels Philosphie des Rechtes." In *Hegels Philosophie des Rechts*, ed. Dieter Henrich and Rolf-Peter Horstmann, 317–81. Stuttgart: Klett Cotta, 1982.

Vető, Miklos. "Le trois images de l'absolu." *Archives de Philosophie* 55 (1992).

Weber, Samuel. *Mass Mediasaurus: Form, Media, Technics*. Ed. Alan Cholodenko. Stanford: Stanford University Press, 1996.

Weischedel, Wilhem. *Der frühe Fichte: Aufbruch der Freiheit der Gemeinschaft*. Stuttgart: Frommann-Holzboog, 1973.

Williams, Robert R. "Hegel's Concept of Geist." In *Hegel's Philosophy of Spirit*, ed. Peter Stillman, 1–20. Albany: SUNY Press, 1986.

————. *Recognition: Fichte and Hegel on the Other*. Albany: SUNY Press, 1992.

Wurzer, Wilhelm. "Fichte's Paragonal Visibility." In *Fichte: Historical Contexts/Contemporary Controversies*, ed. Daniel Breazeale and Tom Rockmore. Atlantic Highlands, N.J.: Humanities Press, 1994.

Zöller, Günter. *Fichte's Transcendental Philosophy*. New York: Cambridge University Press, 1998.

Index